KUMON MATH WORKBOOKS

Word Problems

Table of Contents

1 Maggie is at a jewelry warehouse looking at materials for her jewelry. If she wants 1.5 meters of gold-plated braid, which costs $200 per meter, how much will she have to pay? 10 points

〈**Ans.**〉

2 In art class, we have a 3.8-meter piece of tape. If we cut it into 0.5-meter pieces, how many pieces will we get, and how much will we have left? 10 points

〈**Ans.**〉

3 Mrs. Hazan has two ribbons for wrapping her presents. The white ribbon is $\frac{5}{7}$ meter long, and it is $\frac{1}{7}$ meter longer than the red ribbon. How long is the red ribbon? 10 points

〈**Ans.**〉

4 Gayle's father is trying to clean up his garage. He found a can with $\frac{4}{5}$ liter of oil, and then another can with $\frac{1}{5}$ liter of oil. If he pours them into one can, how many liters of oil will he have? 10 points

〈**Ans.**〉

5 How many times bigger is a 9-square-foot flowerbed than an 8-square-foot one? Answer in a fraction. 10 points

〈**Ans.**〉

6 Josh's older brother is 162 centimeters tall. If he is 1.2 times bigger than Josh, how tall is Josh?

10 points

⟨Ans.⟩

7 The population in Clare's city is made up of 62,950 males and 58,327 females. How many people live in her city in all? Calculate after rounding to the nearest thousands place.

10 points

⟨Ans.⟩

8 Kei's class has 35 students. If 14 are boys, what is the ratio of boys to the total students in Kei's class? Answer in decimal form.

10 points

⟨Ans.⟩

9 In science class, we made salt water with 200 ounces of water and 50 ounces of salt. What is the percentage of salt weight in the weight of the whole?

10 points

⟨Ans.⟩

10 Calculate the percentages in order to fill the table on the right.

10 points for completion

Living zone populations

Zone	Number of people	Percentage
A	24	( )
B	16	( )
C	6	( )
D	4	( )
Total	50	( )

Remember all this? Good!

2 Review

Level ☆

Date / /

Name

Score /100

1 My grandfather's car gets 8 miles to the liter. How far can he go with 8.5 liters of gas? 10 points

〈**Ans.**〉

2 Kim is shopping for the materials for a dress she wants to make. She bought 1.5 meters of red cloth for $12. How much did she pay per meter? 10 points

〈**Ans.**〉

3 We dried 1 kilogram of seawater and got 32.8 grams of salt. If we dried 30 kilograms of seawater, how much salt would we get? 10 points

〈**Ans.**〉

4 Chef Henry is making a sauce for one of his desserts. The pot and the sugar inside weighs $\frac{7}{9}$ pound. If the pot weighs $\frac{2}{9}$ pound, how much sugar is inside? 10 points

〈**Ans.**〉

5 Rob went to the red licorice factory and came home with 14.4 yards of red licorice candy. If he divides it up among 4 people, how long will each person's piece be? 10 points

〈**Ans.**〉

6 Our scout troop is trying to decide between two hikes. One is 7 kilometers long and the other is 10 kilometers long. How many times longer is the 10-kilometer hike? Answer in fraction form. 10 points

〈Ans.〉

7 Captain Kassing's ship can hold 350 people. If there are 280 people already on board, what percentage of the capacity is currently on board? 10 points

〈Ans.〉

8 One pound of potatoes has 0.8 pound of water weight. If there are 20 pounds of potatoes, how much water weight is included? 10 points

〈Ans.〉

9 Cheryl bought a skirt for $27. If the original price was $30, what percentage discount did she get off of the original price? 10 points

〈Ans.〉

10 Peter runs the ice cream section of the local grocery store. If he buys one ice cream container for $5, and he wants a 20% profit, how much should he sell the container for? 10 points

〈Ans.〉

Alright, let's get started!

3 Fractions

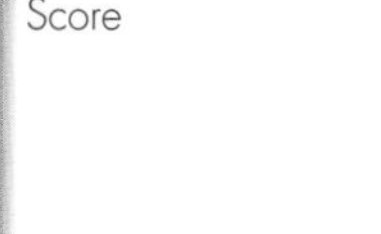

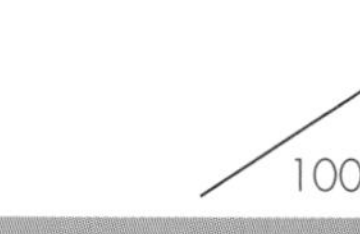

1 Mr. Rostkowski is always trying to clean up his garage. He has $\frac{1}{2}$ liter of oil in one can and $\frac{1}{3}$ liter in another. How much will he have if he pours them both into one can?

10 points

〈Ans.〉

2 Henry's class has a small flowerbed. They planted tulips in $\frac{2}{5}$ square foot and sunflowers in $\frac{1}{4}$ square foot. How many square feet of flowers do they have in all?

10 points

〈Ans.〉

3 I am painting the doghouse with my little brother. I painted $\frac{5}{12}$ of the doghouse, and my little brother painted $\frac{2}{9}$. How much of the doghouse did we finish painting?

10 points

〈Ans.〉

4 Hanna used $\frac{3}{5}$ meter of blue ribbon and she still has $\frac{3}{8}$ meter. How much ribbon did she have at first?

10 points

〈Ans.〉

5 Julie went shopping with her mother and bought $\frac{3}{5}$ pound of oranges. If she put the oranges in a bag that weighed $\frac{1}{4}$ pound, how much did the whole bag weigh?

10 points

〈Ans.〉

6 Mrs. Thomas used $\frac{2}{5}$ liter of oil this week while cooking. If she still has $\frac{3}{8}$ liter, how much oil did she have before this week? 10 points

〈Ans.〉

7 We have $\frac{5}{6}$ pound of black beans and $\frac{3}{4}$ pound of lima beans in a bag. How many pounds of beans do we have? 10 points

〈Ans.〉

8 The distance from Eduardo's house to his school is $\frac{1}{2}$ kilometer. The distance from his school to the train station is $\frac{2}{3}$ kilometer. If he goes past his school, how far is it from Eduardo's house to the train station? 10 points

〈Ans.〉

9 Robin drank $\frac{4}{5}$ pint of milk and his sister drank $\frac{1}{2}$ pint. How much milk did they drink in all? 10 points

〈Ans.〉

10 Evan is almost done decorating the float for the parade. He has $\frac{3}{4}$ meter of white ribbon left and $\frac{2}{5}$ meter red ribbon left. How much ribbon does he have left altogether? 10 points

〈Ans.〉

Fractions? No problem!

4 Fractions

Date / /

Name

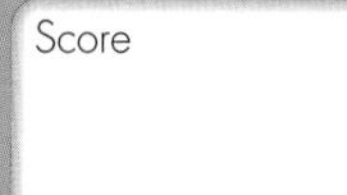

1 Paul was supposed to cook for the whole extended family when they came over for the party. He used $\frac{1}{4}$ liter of soy sauce for one dish. If he had $\frac{2}{3}$ liter when he started, how much soy sauce did he have left?

10 points

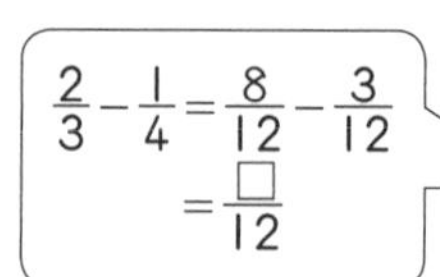

〈Ans.〉

2 In our fridge we have $\frac{2}{3}$ liter of soda and $\frac{1}{2}$ liter of juice. How much more soda than juice do we have?

10 points

〈Ans.〉

3 $\frac{4}{5}$ of the people who rode the train used the pre-paid card, and $\frac{2}{3}$ of the people on the bus did as well. How many more people on the train used the pre-paid card than people on the bus?

10 points

〈Ans.〉

4 Rob is trying to grow taller so he's drinking milk every day. He drank $\frac{2}{7}$ liter yesterday and $\frac{3}{5}$ liter today. Which day did he drink more? How much more?

10 points

〈Ans.〉

5 $\frac{4}{5}$ mile east of Ronda's house is a corner store. $\frac{2}{3}$ mile west of her house is the bakery. Which is further from her house? How much further?

10 points

〈Ans.〉

6 Tim's little sister is mad because he drank so much soda. There was $\frac{8}{9}$ liter of soda this morning, and Tim says he drank only $\frac{1}{3}$ liter. How much soda is left? 10 points

〈**Ans.**〉

7 Carlos is cooking for a party and used $\frac{3}{4}$ pound of rice today. If he had 2 pounds to begin with, how much rice does he have left? 10 points

〈**Ans.**〉

8 The baker used $\frac{1}{3}$ liter of milk in her first cake today. If she had $\frac{1}{2}$ liter of milk to begin with, how much milk does she have left? 10 points

〈**Ans.**〉

9 Jamal is trying to get to the cable car station, which is $\frac{7}{10}$ mile away from where he parked his car. He's already walked $\frac{2}{5}$ mile. How much further does he have to walk to the cable car station? 10 points

〈**Ans.**〉

10 Debby got in trouble and had to paint the fence as her punishment. She painted $\frac{2}{5}$ of the fence and then took a break for lunch. After lunch, she painted $\frac{1}{4}$ of the fence. How much of the fence does she have left to paint? 10 points

Treat the whole fence as 1.

〈**Ans.**〉

You can handle this. Good job!

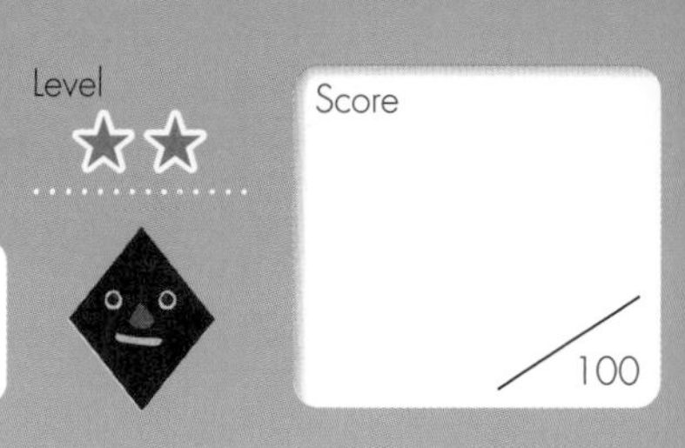

1 Mr. Harper owns a fleet of trucks and is getting ready for the day. It snowed last night so he has a lot to do. He has 3-kilogram bags of salt to give to each of his drivers. If he has 5 bags, how much salt does he have in all?

10 points

Weight per bag		Number of bags		Total weight
3	×	5	=	15

⟨**Ans.**⟩

2 He also has some smaller bags that have $\frac{3}{4}$ kilogram of salt each. If he has 5 of the smaller bags, how much salt does he have?

10 points

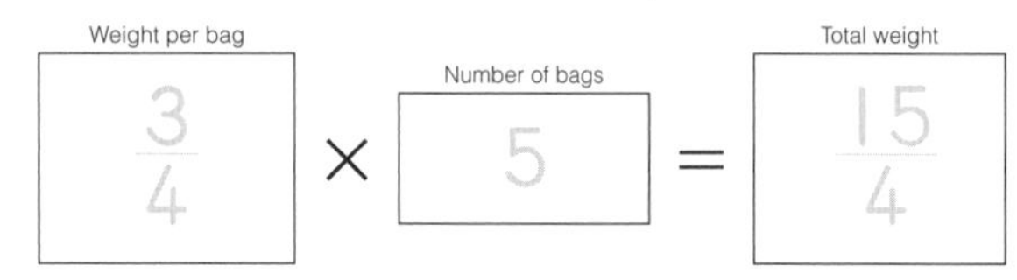

⟨**Ans.**⟩

3 He noticed that some of the trucks need a little fixing up before they can go out. He has some wire to help fix the trucks. If he makes 7 equal-sized pieces of wire that are each $\frac{1}{6}$ yard long, how long much wire did he have at first?

10 points

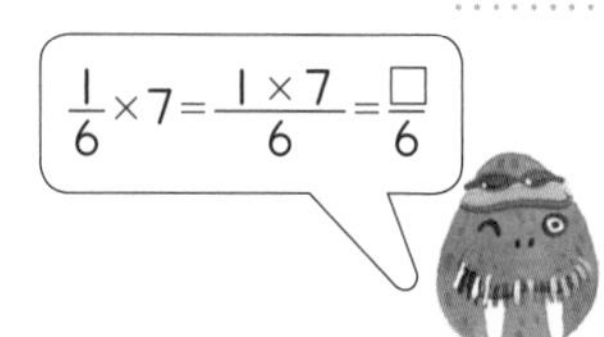

⟨**Ans.**⟩

4 Mr. Harper wants all his drivers to have a snack while they are working, so he bought $\frac{3}{5}$ kilogram of fruit salad for each of them. If he has 6 drivers, how much fruit salad did he bring today?

10 points

⟨**Ans.**⟩

5 They need coffee, too, so Mr. Harper got 3 pitchers that each have $\frac{2}{5}$ liter of coffee in them. How much coffee did he buy in all?

10 points

⟨**Ans.**⟩

6 Mr. Versa is making a birdfeeder for his backyard. He has 5 ounces of paint, and he can paint $\frac{2}{3}$ square meter of birdfeeder with 1 ounce. How much can he paint with all 5 ounces?

10 points

〈Ans.〉

7 Reena thinks she is getting sick and has been drinking orange juice all week for the vitamin C. If she drinks $\frac{3}{4}$ liter of orange juice every day for 6 days, how much orange juice will she drink in all?

10 points

$\frac{3}{4}\times 6=\frac{3\times \overset{3}{\cancel{6}}}{\underset{2}{\cancel{4}}}=\frac{\square}{\square}$

〈Ans.〉

8 Our pen for our little rabbit is a square. If each side is $\frac{3}{4}$ yard long, what is the perimeter of our rabbit pen?

10 points

〈Ans.〉

9 Kevin wants to give all 12 children at his birthday party $\frac{5}{6}$ meter of licorice. How much licorice does he need to buy?

10 points

〈Ans.〉

10 The McNeeley family went through their closets and found 10 bags of clothes to give away to charity. If each bag weighs $\frac{4}{5}$ kilogram, how much do all the clothes they found weigh in all?

10 points

〈Ans.〉

You're a superstar!

Fractions

Date / /

Name

1 Greg is supposed to buy a dessert for dinner on the way home. He's thinking of buying 3 watermelons. If the total weight is 12 pounds, how much does each watermelon weigh? 10 points

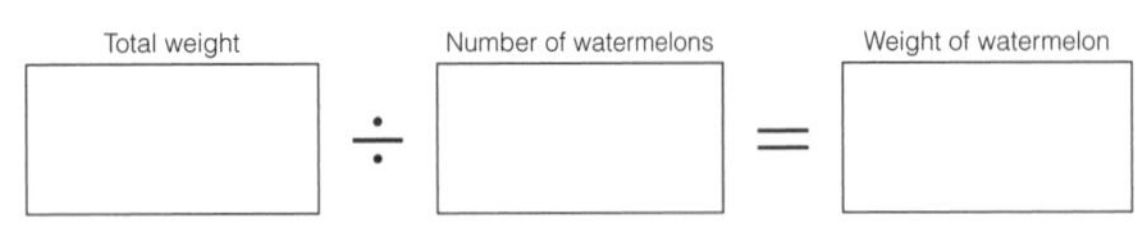

〈Ans.〉

2 Now, Greg is looking at buying 3 tubs of ice cream. If all 3 weigh $\frac{7}{10}$ pound in all, how much does each tub of ice cream weigh? 10 points

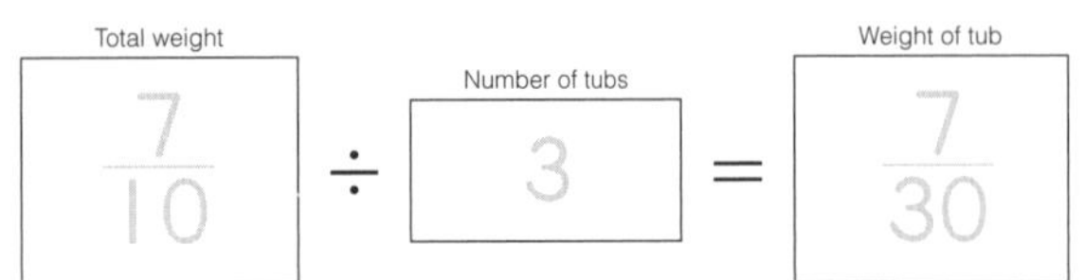

〈Ans.〉

3 The iron bar I work out with weighs $\frac{5}{7}$ kilogram and is 2 yards long. How much would 1 yard of this bar weigh? 10 points

〈Ans.〉

4 Dennis has a bunny, and if it hops 4 times, it goes $\frac{5}{6}$ meter. How far does his bunny go in 1 hop? 10 points

〈Ans.〉

5 Near the end of the hike, Nina and her friends were running out of water. If she split $\frac{4}{5}$ liter of water equally among the 3 of them on the hike, how much water did each person get? 10 points

〈Ans.〉

6 Sterling lives on an island that has one sugar plant. There was an explosion in the plant last month, and there's still no new sugar. When Sterling and his 2 neighbors pooled their sugar, Sterling had $\frac{2}{5}$ pound of sugar. He needed to split it evenly for 3 different recipes. How much sugar did each recipe call for?

10 points

⟨**Ans.**⟩ ____________

$\frac{2}{5} \div 3 = \frac{2}{5} \times \frac{1}{3} = \frac{\square}{\square}$

7 Jefferey just became a milkman this week. He forgot that Mrs. Dickerson likes the small bottles of milk, not the large ones. If he has $\frac{8}{9}$ liter of milk, and he divides it into 4 smaller bottles, how much milk is in each of the smaller bottles?

10 points

⟨**Ans.**⟩ ____________

$\frac{8}{9} \div 4 = \frac{\overset{2}{\cancel{8}}}{9} \times \frac{1}{\underset{1}{\cancel{4}}} = \frac{\square}{\square}$

8 Dr. Kwan is working hard in the lab. He has 3 yards of tubing that weighs $\frac{6}{7}$ pound. How much would 1 yard of the tubing weigh?

10 points

⟨**Ans.**⟩ ____________

9 Bobby's art class is working with clay today. If he divides $\frac{5}{7}$ kilogram of clay among 5 students, how much clay will each student get?

10 points

⟨**Ans.**⟩ ____________

10 Brandon's new toy robot can move $\frac{8}{9}$ meter in 6 seconds. How far did it move per second?

10 points

⟨**Ans.**⟩ ____________

Multiplication and division of fractions? You've got it.

7 Fractions

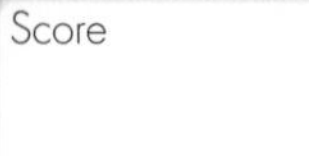

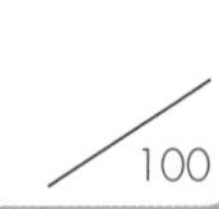

1 The fountain at work flows at the rate of 5 liters of water per minute. How many liters of water flow in 4 minutes? 10 points

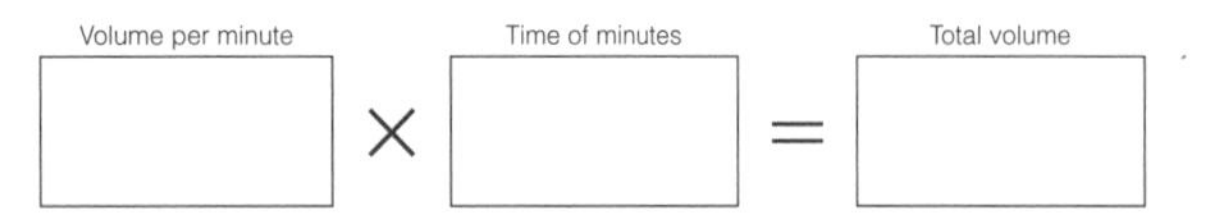

〈Ans.〉

2 They wanted the fountain to be more impressive, so they turned it up, and now the fountain flows at the rate of 7 liters per minute. How many liters of water flow in $\frac{5}{6}$ minute? 10 points

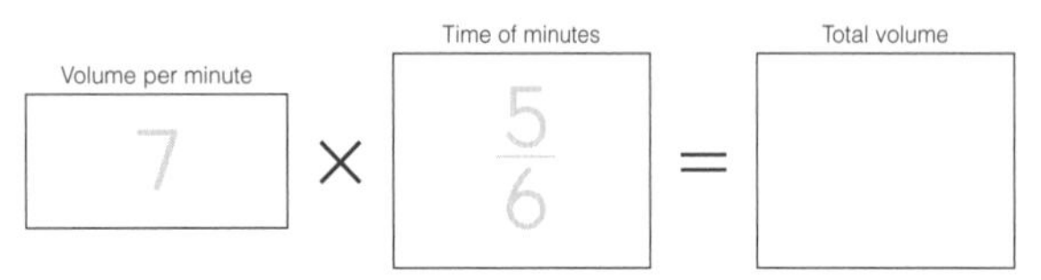

〈Ans.〉

3 Alex can paint 6 square meters of his bedroom with 1 liter of paint. If he uses $\frac{3}{5}$ liter of paint, how much of his room can he paint? 10 points

〈Ans.〉

4 Flo's favorite ribbon costs 80¢ for 1 yard. If she buys $\frac{3}{4}$ yard of ribbon, how much will she spend? 10 points

〈Ans.〉

5 Jordan forgot to turn off the hose. If the hose flows at 9 liters per minute, and the water was on for $\frac{4}{5}$ minute, how much water spilled out while he was gone? 10 points

〈Ans.〉

6 I yard of David's favorite gum tape weighs $\frac{2}{3}$ pound. How much would $\frac{4}{5}$ yard weigh? 10 points

Weight per yard		Length of gum tape		Total weight
 $\frac{2}{3}$	$\times$	$\frac{4}{5}$	$=$	

⟨**Ans.**⟩

7 The electrician knows that I meter of wire weighs $\frac{1}{8}$ kilogram. How much would $\frac{3}{7}$ meter weigh? 10 points

⟨**Ans.**⟩

8 If there is about $\frac{1}{5}$ square yard of weeds in I square yard of our lawn, how much area is covered by weeds in $\frac{3}{4}$ square yard of lawn? 10 points

⟨**Ans.**⟩

9 My poster for school is $\frac{4}{5}$ yard long and $\frac{1}{2}$ yard wide. What is the area of my poster? 10 points

Area = Length × Width

⟨**Ans.**⟩

10 You can paint $\frac{3}{5}$ square yard of the fence with I ounce of paint. How much of the fence can you paint with $\frac{2}{3}$ ounce of paint? 10 points

⟨**Ans.**⟩

Don't forget to reduce to make the problem easier!

Fractions

Date / /

Name

1 In Claire's house, they drank $\frac{5}{6}$ liter of milk this morning. Dylan's house drank twice as much. How much milk did they drink at Dylan's house? 10 points

Claire's house: $\frac{5}{6}$ × How many times more: 2 = Dylan's house: ☐

⟨Ans.⟩ ______

2 Allison got 2 pounds of candy over the holidays from all of her relatives. Brenda got $\frac{3}{4}$ as much. How much candy did Brenda get? 10 points

Allison's candy: 2 × How many times less: $\frac{3}{4}$ = Brenda's candy: ☐

⟨Ans.⟩ ______

3 George threw the discus 36 yards. Ray threw it $\frac{2}{3}$ as far. How far did Ray throw the discus? 10 points

⟨Ans.⟩ ______

4 Pablo can hop $\frac{3}{4}$ meter on one leg. Casey jumped $\frac{5}{6}$ as far. How far did Casey jump? 10 points

$\frac{3}{4} \times \frac{5}{6} =$

⟨Ans.⟩ ______

5 Jim got $\frac{4}{5}$ kilogram of clay in art class today. Steve got $\frac{2}{3}$ times as much. How much clay did Steve get? 10 points

⟨Ans.⟩ ______

6 Gil is trying to decide between red and blue paint for his room. He has $\frac{7}{8}$ liter of red paint, and he has $\frac{3}{4}$ as much blue paint. How much blue paint does he have? 10 points

⟨**Ans.**⟩

7 There are 360 students in Andrea's school. Her brother goes to a different school, which has $\frac{3}{5}$ times as many students. How many students go to her brother's school? 10 points

⟨**Ans.**⟩

8 Chuck and Mark went grape picking. Chuck picked $\frac{6}{7}$ kilogram of grapes. Mark picked $\frac{2}{3}$ of grapes Chuck picked. How much did the grapes that Mark picked weigh? 10 points

⟨**Ans.**⟩

9 Claudia wants to bake a cake. The recipe calls for $\frac{9}{10}$ pound of flour to start. She only has $\frac{1}{3}$ of the flour she needs. How much flour does she have? 10 points

⟨**Ans.**⟩

10 Billy is trying to fix his radio-controlled car and thinks he can use either string or wire to hold a piece in place. He has $\frac{8}{5}$ feet of wire. If he has $\frac{5}{6}$ as much string, how much string does he have? 10 points

⟨**Ans.**⟩

You're doing really great. Nice!

Fractions

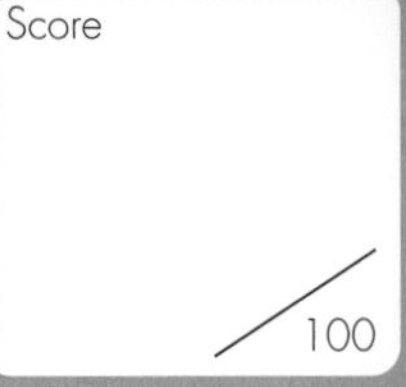

1 Mrs. Reeds is running water for a bath. 10 liters of water is flowing every 3 minutes from her tap. How much water is flowing per minute? 10 points

$10 \div 3 =$

⟨**Ans.**⟩

2 Mrs. Reeds got tired of waiting and turned both faucets up as high as they could go. Now 10 liters of water flows every $\frac{3}{5}$ minute from her tap. How much water is flowing per minute? 10 points

$10 \div \frac{3}{5} = 10 \times \frac{5}{3}$

⟨**Ans.**⟩

3 Joe's mother makes really rich brownies. If she makes 2 brownies with $\frac{1}{3}$ a stick of butter, how many brownies does she make with a whole stick of butter? 10 points

⟨**Ans.**⟩

4 At my grocery store, they have a deal where $\frac{4}{5}$ pound of chestnuts are \$8. How much is that per pound? 10 points

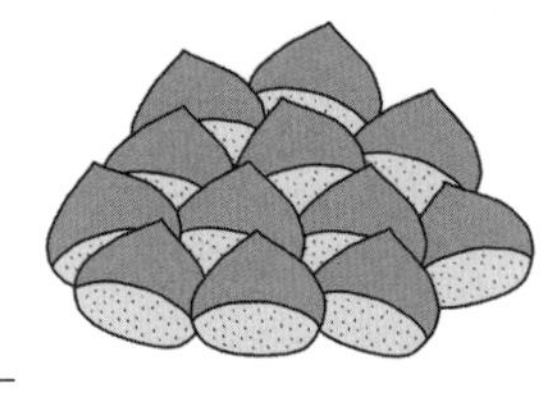

⟨**Ans.**⟩

5 Carlos is baking cookies for the bake sale. $\frac{3}{5}$ liter of oil can make 450 cookies. How many cookies can Carlos bake if he uses 1 liter of oil? 10 points

⟨**Ans.**⟩

6 Victor is playing around with his scale. 2 of his action figures weigh $\frac{5}{8}$ pound. How much does each action figure weigh? 10 points

$\frac{5}{8} \div 2 =$

〈Ans.〉

7 Mr. Walsh had a busted pipe in his basement. He sawed a piece out and weighed it to make sure he knew what kind of pipe to get. If he knows that $\frac{2}{5}$ meter of that pipe weighs $\frac{3}{8}$ kilogram, how much should 1 meter weigh? 10 points

$\frac{3}{8} \div \frac{2}{5} =$

〈Ans.〉

8 Donna is sewing shirts for her dolls. She knows she can sew $\frac{2}{3}$ of one shirt with $\frac{5}{6}$ yard of fabric. How much of a shirt can she sew if she uses 1 yard? 10 points

〈Ans.〉

9 Victor is playing with his bathroom scale again. He fit $\frac{2}{3}$ of his action figures on the scale at once, and they weighed $\frac{7}{10}$ kilogram. How much would all of his action figures weigh? 10 points

〈Ans.〉

10 Farmer Wagner spread $\frac{2}{7}$ pound of fertilizer in a $\frac{5}{6}$ square-yard section of his field. If he uses this same ratio, how much ground can he cover with a full pound of fertilizer? 10 points

〈Ans.〉

Phew. Take a break if you need one!

Fractions

Date / /

Name

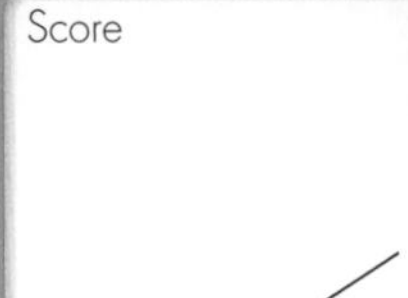

1 Our class decided to play a new version of hide and go seek inside because the weather was bad. Each of us got a different color string and had to split it into pieces and tie the pieces somewhere in the room while the lights were out. If we split up 12 inches of string into pieces that were 2 inches long, how many pieces did we have to hide? 10 points

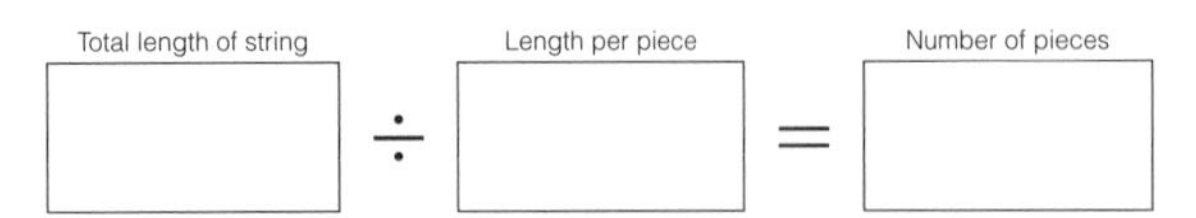

〈**Ans.**〉

2 We decided to make it harder and add more strings. If we cut 7 inches of string into pieces that were $\frac{1}{6}$ inch long, how many pieces could we make? 10 points

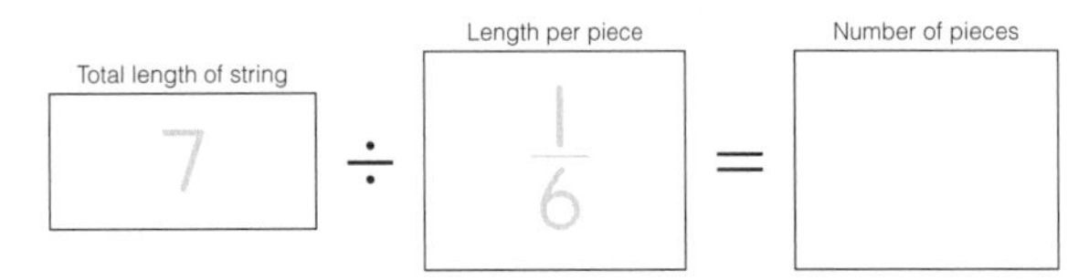

〈**Ans.**〉

3 After we finished playing, we divided up the cookies that Mrs. Jones cooked. If there were 5 plates of cookies, and we each got $\frac{1}{4}$ of a plate, how many of us got cookies? 10 points

〈**Ans.**〉

4 Mrs. Sandel is sewing ribbons on some of her daughter's dolls. If she cuts a 6-foot ribbon into pieces that are $\frac{6}{5}$ feet long, how many pieces will she make? 10 points

〈**Ans.**〉

5 Kori and Sander decided to have a soda-drinking competition. They have 4 liters of soda. If they put $\frac{2}{3}$ liter in each cup, how many cups will they need? 10 points

〈**Ans.**〉

6 Mr. Shiller wants his boys to help salt the neighborhood after the blizzard. He divided $\frac{3}{4}$ kilogram of salt into bags that could hold $\frac{1}{12}$ kilogram each. How many bags of salt did he make? 10 points

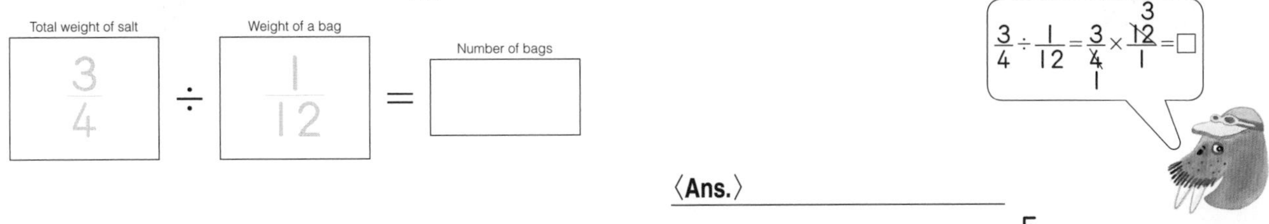

〈**Ans.**〉

7 Ramon is packing a present to send to his friend. He has a piece of tape that is $\frac{5}{6}$ foot long. If he cuts it into pieces that are $\frac{5}{12}$ foot long each, how many pieces will he make? 10 points

〈**Ans.**〉

8 Mrs. Jacobson made a long lemon bar for dessert. If it was $\frac{2}{3}$ foot long, and she cut it into pieces that were $\frac{2}{15}$ foot long, how many pieces did she make? 10 points

〈**Ans.**〉

9 Our study group each decided to bring a snack to our study session. I brought $\frac{4}{5}$ liter of juice. If we put $\frac{4}{25}$ liter in each cup, how many cups of juice will we have? 10 points

〈**Ans.**〉

10 There was a hole in our $\frac{5}{4}$-kilogram bag of sugar in the pantry. We decided to break the bag into smaller bags with $\frac{5}{8}$ kilogram in each bag. How many new bags of sugar will we have? 10 points

〈**Ans.**〉

Always remember to reduce.
It makes the problems easier!

1 The Gorman sisters are both learning to knit. Sarah got 8 feet of yarn to start with, while Monica got 2 feet. How many times more yarn did Sarah get than Monica? 10 points

Sarah $\square$ ÷ Monica $\square$ = How many times more $\square$

⟨**Ans.**⟩

2 Mrs. Gorman felt bad that she gave Monica so much less yarn. The next time they knitted, she gave Sarah $\frac{3}{5}$ foot of yarn and gave Monica 1 foot. How many times more yarn did Monica get than Sarah? 10 points

Monica $\square$ ÷ Sarah $\square$ = How many times more $\square$

⟨**Ans.**⟩

3 We're working with clay in art class today. We have 4 pounds of white clay and $\frac{4}{5}$ pound of green clay. How many times more white clay do we have than green clay? 10 points

⟨**Ans.**⟩

4 John built an area in the backyard for the dog because he felt bad that the dog could only go outside when John had time take him on a walk. He built a square pen that was 8 square feet big, and then a little rectangular bed that was $\frac{8}{9}$ square foot. How many times larger was the pen than the bed? 10 points

⟨**Ans.**⟩

5 Julia helped her father mow the yard. He mowed 14 square feet and she mowed $\frac{7}{8}$ square foot. How many times more area did Julia's father mow? 10 points

⟨**Ans.**⟩

6 The Gordons got a baby rabbit and a kitten over the holidays. The baby rabbit weighs $\frac{4}{5}$ kilogram and the kitten weighs $\frac{3}{5}$ kilogram. How many times heavier is the baby rabbit than the kitten?

10 points

Rabbit's weight		Kitten's weight		How many times more
	$\div$		$=$	

〈Ans.〉

7 Kyle's younger sister is always comparing what she has to what Kyle has. Today, she has $\frac{6}{7}$ liter of juice while Kyle has $\frac{15}{14}$ liters of juice. How many times more juice does Kyles younger sister have?

10 points

Answer with reduced fractions.

〈Ans.〉

8 Alicia is making a fruit salad. She has $\frac{4}{5}$ pound of grapes and $\frac{2}{3}$ pound of apples. How many times more apples does she have than grapes?

10 points

〈Ans.〉

9 Greg framed his favorite picture. The square picture was $\frac{3}{8}$ square foot and the rectangular frame was $\frac{9}{8}$ square feet. How many times smaller was the picture?

10 points

〈Ans.〉

10 Felipe painted a wall of the garage with his brother because their father asked them to. Felipe painted $\frac{2}{3}$ square yard, while his brother painted $\frac{7}{9}$ square yard before they got tired. How many times more area did his brother paint?

10 points

〈Ans.〉

Keep on plugging!

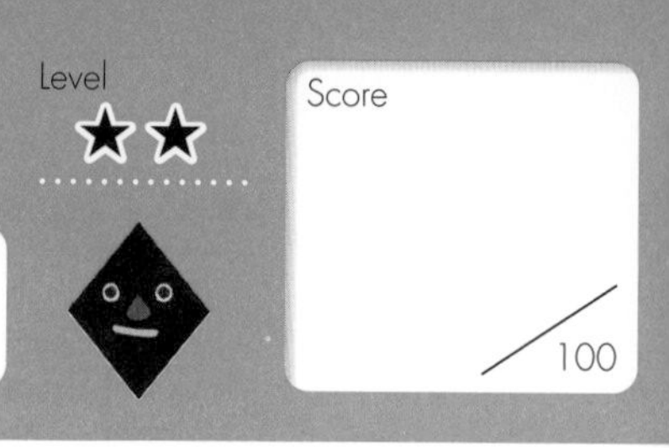

1 Kate needed 3 pieces of purple yarn for a glove she was making. Each piece had to be $\frac{4}{5}$ meter. If she bought 5 meters of yarn, how much yarn did she have left over? 10 points

$5-3\times\frac{4}{5}=$

〈Ans.〉

2 Kate also wanted to make a scarf. She needed 2 pieces of gold yarn and each piece had to be $\frac{3}{4}$ meter. If she bought 5 meters of yarn, how much yarn did she have left over? 10 points

〈Ans.〉

3 Ray's favorite gum comes in tape form. He had 5 feet of gum tape and cut 4 pieces that were each $\frac{5}{6}$ foot long. How much gum did he have remaining? 10 points

〈Ans.〉

4 Ted's father left 1 liter of juice in the fridge. Then Ted and his brother came and drank $\frac{2}{5}$ liter each. How much juice did they leave for their father? 10 points

〈Ans.〉

5 The wire we use in art class weighs $\frac{1}{5}$ pound per yard. We had 6 yards, and then used 1 yard for our dioramas. How much does the remaining wire weigh? 10 points

〈Ans.〉

6 Hugo wanted to pick up some food for dinner. He got 4 large containers of salad and 6 smaller containers of pasta. If the larger containers held $\frac{7}{10}$ pound of salad, and the smaller containers held $\frac{3}{5}$ pound of pasta, how much did all containers weight together? 10 points

〈**Ans.**〉

7 Colin always takes his laundry to the laundromat for cleaning. Today, he has 5 items for dry cleaning and 12 items for regular washing. If each item of dry cleaning weighs $\frac{3}{4}$ pound, and each item of regular laundry weighs $\frac{5}{16}$ pound, how much does all of Colin's laundry weigh? 10 points

〈**Ans.**〉

8 Dave's father is a construction worker and needs to know the weight of two iron bars. The 20 foot iron bar lying by the entrance of the contruction site weighed $\frac{3}{4}$ pound per foot. The 10 foot iron bar by the back was $\frac{3}{5}$ pound per foot. How much do both bars weigh together? 10 points

〈**Ans.**〉

9 Jane needs 180 centimeters of green cloth and 240 centimeters of red cloth to make her holiday wreath. If she only had the money to buy $\frac{5}{6}$ of each length that she needed, how many centimeters of cloth did she bring home in all? 10 points

〈**Ans.**〉

10 Sam was painting his art project, and one side looked like the picture below. How much area was on that side of his project? 10 points

$\frac{5}{6}$ ft.

$\frac{4}{5}$ ft. $\frac{3}{5}$ ft.

〈**Ans.**〉

Okay, time to change it up a little.
Are you ready?

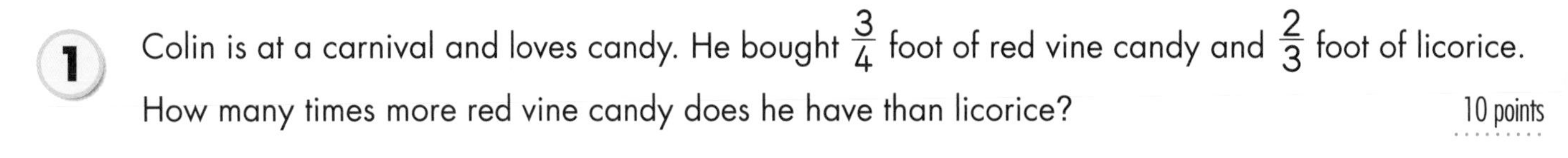

1. Colin is at a carnival and loves candy. He bought $\frac{3}{4}$ foot of red vine candy and $\frac{2}{3}$ foot of licorice. How many times more red vine candy does he have than licorice? 10 points

〈Ans.〉

2. Tiana and Toni went to get their hair cut. They both decided to cut a lot of hair. Tiana got $\frac{3}{4}$ foot cut off, and Toni cut $\frac{2}{3}$ foot of hair off. What is the ratio of the length of hair Tiana cut to the length of hair Toni cut? 10 points

〈Ans.〉

3. What is the ratio of the width to the length in the rectangle pictured below? 10 points

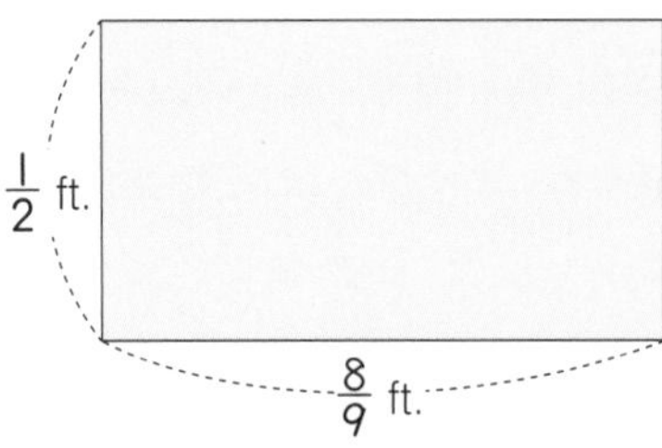

〈Ans.〉

4. Prudence has a dog named Beanie. They had $\frac{4}{5}$ pound of dog food, but Beanie got into the dog food and ate $\frac{2}{5}$ pound. What is the ratio of the dog food remaining to the original amount of dog food? 10 points

Weight of remaining dog food: $\left(\frac{4}{5} - \frac{2}{5}\right)$ ÷ Weight of dog food at first: $\frac{4}{5}$ = Ratio: ☐

〈Ans.〉

5. Mr. Kemp is building a fence around his backyard. He has a piece of wood that is $\frac{5}{6}$ yard long, and he cut $\frac{1}{6}$ yard off. What is the ratio of the length of the wood remaining to the original length? 10 points

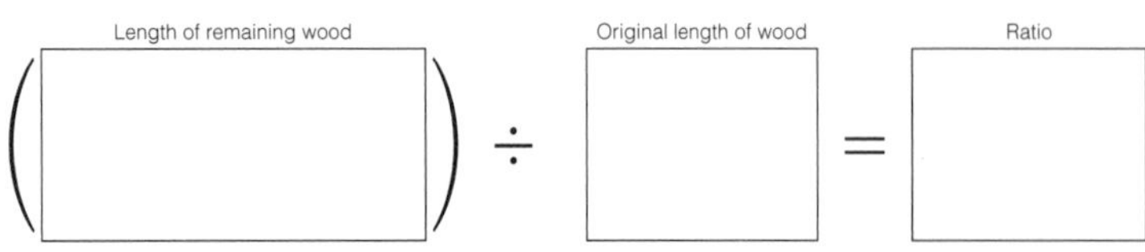

〈Ans.〉

6 Gina had $\frac{3}{4}$ liter of juice on the table. When she turned around, her little sister knocked the juice over by accident. Gina lost $\frac{1}{4}$ liter of juice. What is the ratio of the juice remaining to the juice she had at first?

10 points

⟨Ans.⟩

7 Keith and Randy had a water drinking contest. Keith drank $\frac{2}{5}$ liter of water without taking a breath, and Randy finished $\frac{1}{3}$ liter. What is the ratio of the amount of water Keith drank out of the total amount of water they both drank?

10 points

Amount Keith drank $\square$ ÷ (Total amount boys drank $\square$) = Ratio $\square$

⟨Ans.⟩

8 Mrs. Brown made a pie, and her sons Aaron and Chris ate most of it. Aaron ate $\frac{1}{2}$ of the pie, and Chris ate $\frac{1}{3}$. What is the ratio of the amount Aaron ate to the amount both brothers ate?

10 points

⟨Ans.⟩

9 Naomi needs some soy sauce for her recipe. She found $\frac{3}{5}$ cup left in the big bottle, and another $\frac{1}{4}$ cup in a small bottle. What is the ratio of the amount of soy sauce in the small bottle to the total amount of soy sauce she has left?

10 points

⟨Ans.⟩

10 Jim needs a whole bell pepper for his salad. If he has $\frac{3}{8}$ of a yellow bell pepper, and $\frac{1}{4}$ of a green bell pepper, what is the ratio of the green bell pepper to his total amount of bell pepper?

10 points

⟨Ans.⟩

You're getting this, right? Good!

Ratios

Date / /

Name

Level ★★

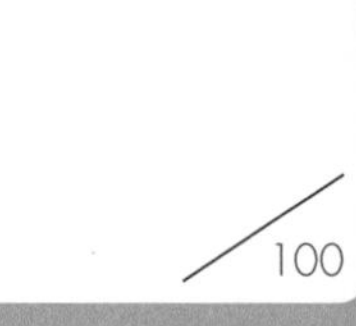

Score /100

1 There are 120 people on the train to Littleton. Men make up $\frac{3}{4}$ of the whole. How many men are there on the train? 10 points

⟨Ans.⟩

2 Anne likes a mix of apple juice and lemonade. Her mother made her 2 liters of the mix, and $\frac{3}{4}$ of that amount was lemonade. How much lemonade did her mother use? 10 points

⟨Ans.⟩

3 Tammy is reading a book about wizards that is 240 pages long. She finished $\frac{3}{8}$ of the book today. How many pages did she read? 10 points

⟨Ans.⟩

4 We have two types of soda in the fridge to offer to guests. We have $\frac{3}{4}$ liter of cola. If we have $\frac{2}{3}$ as much ginger ale as cola, how much ginger ale do we have? 10 points

$\frac{3}{4} \times \frac{2}{3} =$

⟨Ans.⟩

5 Rebecca can hop $\frac{4}{5}$ yard on one foot. Her sister can only hop half as far. How far can her sister hop? 10 points

⟨Ans.⟩

6 If there are 90 boys in sixth grade at my school, and that represents $\frac{3}{4}$ of all of sixth grade, how many sixth graders are there at my school? 10 points

⟨**Ans.**⟩

7 Mrs. Lee made pies for a pie-eating contest at the picnic. Her son Eddie ate 3 pies, and that was $\frac{3}{4}$ of the amount that his brother David ate. How many pies did David eat? 10 points

⟨**Ans.**⟩

8 Eli is 144 centimeters tall. That is $\frac{9}{10}$ as tall as his older brother. How tall is Eli's older brother? 10 points

⟨**Ans.**⟩

9 Mr. Snyder is trying to decide about dinner and is looking at the potatoes and the rice in the pantry. He has $\frac{4}{7}$ pound of potatoes, and that is $\frac{4}{5}$ the amount of rice. How much rice does Mr. Snyder have? 10 points

$\frac{4}{7} \div \frac{4}{5} =$

⟨**Ans.**⟩

10 Mrs. Levin has two pitchers, but one is cracked near the top. She poured the lemonade from the cracked pitcher into the better pitcher before the guests came. If she poured in $\frac{8}{9}$ liter of lemonade, and that is $\frac{2}{3}$ of the total volume of the new pitcher, how much can the new pitcher hold in all? 10 points

⟨**Ans.**⟩

Good job. Now let's try something different.

1 The chart on the right shows the number of books Henry reads per month from September to December. What was the average amount of books he read per month? 10 points

Number of books Henry read per month

September	October	November	December
3	5	2	6

(Total number of books: 3 + 5 + 2 + 6) ÷ Number of months: 4 = Average: ____

⟨**Ans.**⟩

2 Tricia wondered if all the eggs in her fridge weighed a similar amount. She weighed 5 eggs and got the results below. What was the average weight? 10 points

3.1 oz　　2.9 oz　　3.2 oz　　3 oz　　2.8 oz

⟨**Ans.**⟩

3 Rick was having trouble in math class. His teacher told him to retake a certain test until he could get a good score. He got a 75 twice, an 80 twice, and finally he got a 94. What was his average score? 10 points

⟨**Ans.**⟩

4 Tim, George and Sam compared the weights of their little brothers. Tim's little brother weighs 64.8 pounds, George's weighs 57.6 pounds, and Sam's weighs 77.4 pounds. What is the average weight of their little brothers? 10 points

⟨**Ans.**⟩

5 Nika was bored, so she decided to measure her friends to see how tall they were. She measured 136 centimeters, 140 centimeters, 145 centimeters, 132 centimeters, and 141 centimeters. What was the average height? Round to the nearest whole number. 10 points

⟨**Ans.**⟩

Don't forget!

The **mean** is the average of a set of numbers.

Average = mean = sum of set of numbers ÷ amount of numbers in set

6 The table below shows the number of students absent from sixth grade last week. What was the average amount of students that were absent per day? 10 points

Absent students in sixth grade

Mon.	Tue.	Wed.	Thu.	Fri.
2	1	0	3	1

〈**Ans.**〉

7 Maria charted how many eggs her family ate last week. How many eggs did they eat on average per day? Round your answer to the nearest tenth. 10 points

Number eggs eaten by Maria's family

Mon.	Tue.	Wed.	Thu.	Fri.	Sat.	Sun.
4	3	1	0	4	4	2

〈**Ans.**〉

8 Leslie's class split up into groups and measured a certain hall at school. The result from Group A was 63.5 meters, Group B had 62.8 meters, Group C had 63.4 meters, and Group D had 63.6 meters. What was the average result? Round your answer to the nearest tenth. 15 points

〈**Ans.**〉

9 Fabio got sick, and while he was home, he watched the construction site across the road. For the first three days, 5 trucks came to the site every day. The next four days, 4 trucks came to the site each day. How many trucks came to the site on average each day? Round your answer to the nearest tenth. 15 points

〈**Ans.**〉

You´re way above average. Way to go!

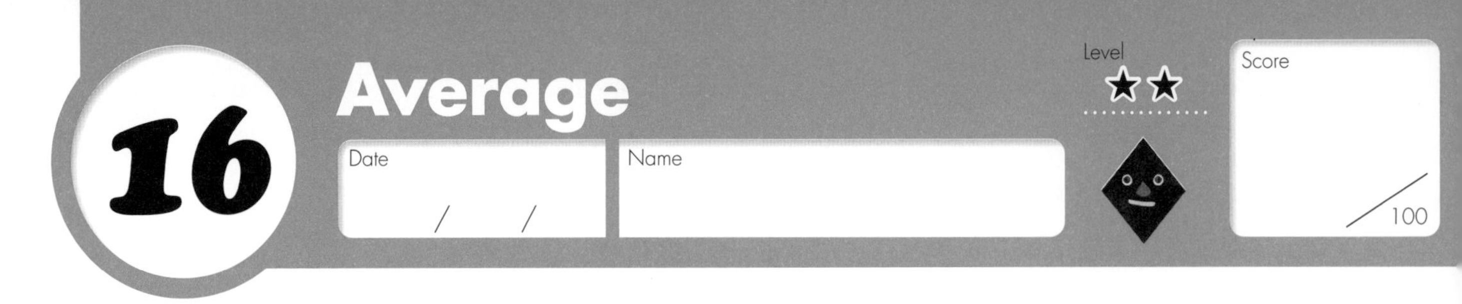

1. There are 6 people in Sara's book club. If they each read 4 books on average in September, how many books did the book club read in all? 10 points

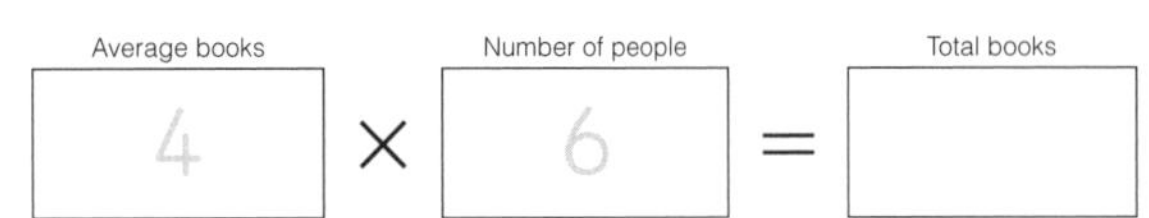

〈Ans.〉

2. May is trying to read a little bit every day. If she read, on average, 36 pages a day for 7 days, how many pages did she read? 10 points

Average pages □ × Number of days □ = Total pages □

〈Ans.〉

3. Tony and his family went fishing. The 4 of them caught an average of 4.5 fish each. How many fish did his family catch? 10 points

〈Ans.〉

4. This September, Ian's class averaged 0.8 of an absent student per day. If they had 20 school days in September, how many students were absent in all? 10 points

〈Ans.〉

5. Audrey measured her stride and found out it was 0.65 yard on average. If it took her 600 steps to get to school, approximately how far is it from her house to school? 10 points

〈Ans.〉

6 Rosa is having a tough time in French class. She has averaged 75 points on her last 3 tests. If she gets a 95 on the fourth test, what will her average be? Answer each step below. 5 points per question

(1) What was her total score for the first 3 tests?

〈Ans.〉

(2) What would her total score for 4 tests be?

〈Ans.〉

(3) What is her average score for all 4 tests?

〈Ans.〉

7 Matt's 4 friends average 4.6 feet tall. If Matt is 4.1 feet tall, what is the average for all 5 friends? 10 points

〈Ans.〉

8 Lisa has been doing very well on her spelling tests. She has taken 4 so far this year and her average is 88. What score will she have to get on her fifth test in order to average 90? Answer each step below. 5 points per question

(1) What would her total score be if she averaged 90 over 5 tests?

〈Ans.〉

(2) What was her total score for the first 4 tests?

〈Ans.〉

(3) What score will she need on her fifth test in order to average 90?

〈Ans.〉

9 Maria was weighing her eggs. She weighed 5 eggs and got an average weight of 62.5 grams. When she added the sixth egg, her average was 62 grams. How much did the sixth egg weigh? 10 points

〈Ans.〉

Okay, now you're ready to try something tougher!

Quantity per Unit

Level 

Date / /

Name

Score 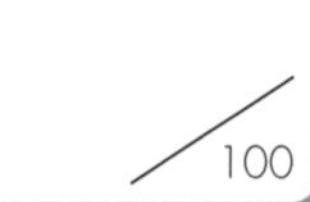

/100

1 In order to answer the questions below, use the illustrations of the 3 chicken coops pictured here.

8 points per question

A 

Area of coop **A** : 5 m^2

B 

Area of coop **B** : 5 m^2

C 

Area of coop **C** : 6 m^2

(1) Which coop is more crowded, **A** or **B**? ( )

(2) Which coop is more crowded, **B** or **C**? ( )

(3) Find the number of chickens per 1 square meter in coops **A** and **C**. Round your answers to the nearest tenth.

(A) $6 \div 5 =$ 〈**Ans.**〉( )

(C) 〈**Ans.**〉( )

(4) Which is more crowded, a coop with a larger or smaller number per square meters?

( )

(5) Find how much area (m^2) each chicken has in coops **A** and **C**. Round your answers to the nearest hundredth.

(A) $5 \div 6 =$ 〈**Ans.**〉( ) m^2

(C) 〈**Ans.**〉( ) m^2

(6) Which is more crowded, a coop with a larger or smaller area (m^2) per chicken?

( )

(7) Which coop is more crowded, **A** or **C**?

( )

2 The table on the right shows the number of people playing at parks A and B, and how much space they have.

8 points per question

Number of people playing at two parks and their areas

	Area (m^2)	Number of people
Park A	140	56
Park B	200	90

(1) Find the number of people playing per 1 m^2 in each park.

(Park A) $56 \div 140 =$

〈Ans.〉 (　　　　)

(Park B)

〈Ans.〉 (　　　　)

(2) Find the area per person in each park. Round your answer to the nearest tenth.

(Park A) $140 \div 56 =$ 〈Ans.〉 (　　　　)

(Park B) 〈Ans.〉 (　　　　)

(3) Which park is more crowded? (　　　　)

Don't forget!

The population per unit of area is called **population density**.

3 The table on the right shows the population and area for Towns A and B.

10 points per question

	Area (km^2)	Population
Town A	4,144	820,000
Town B	5,676	1,490,000

(1) Find the population density of each town. Round the area to the nearest hundreds place before calculating.

(Town A) $820{,}000 \div 4{,}100 =$

〈Ans.〉 (　　　　)

(Town B)

〈Ans.〉 (　　　　)

(2) Which town has a higher population density? (　　　　)

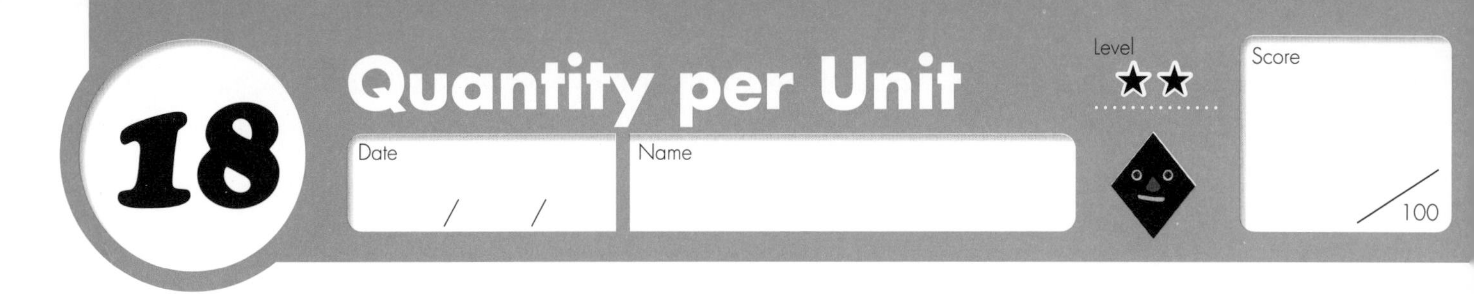

18 Quantity per Unit

Level ★★

Date / / Name

Score /100

1 Farmer Harris always competes with Farmer Wood. Yesterday, they harvested 90 kilograms of potatoes from 50 square meters at the Harris farm. At the Wood farm, they harvested 57 kilograms from 30 square meters. Which farm harvested more potatoes per square meter? 10 points

Harris farm potatoes kg / m² $90 \div 50 = 1.8$

Wood farm potatoes kg / m² $57 \div 30 = 1.9$

〈**Ans.**〉

2 Car A can go 315 kilometers with 35 liters of gas. Car B can go 380 kilometers with 40 liters of gas. Which car is more fuel efficient (which car can go further per liter)? 10 points

Car A km / L $315 \div 35 =$

Car B km / L $380 \div 40 =$

〈**Ans.**〉

3 Helen is shopping for dinner and can't decide between the red potatoes and the regular potatoes. The red potatoes are $3 for 1.5 pounds and the regular ones are $4 for 1.6 pounds. Which potato is more expensive? 10 points

Red potato price per pound

Regular potato price per pound

〈**Ans.**〉

4 The red ribbon at the fabric store is $1.50 for 6 feet and the white ribbon is $1.30 for 5 feet. Which ribbon is cheaper? Find each price per foot first. 10 points

Red ribbon price per foot

White ribbon price per foot

〈**Ans.**〉

5 Cindy is shopping for school. There are 5 green notebooks for $6.25 and 4 brown notebooks for $5.20. Which notebooks are cheaper? Find the price per notebook first. 10 points

Green notebooks price per book

Brown notebooks price per book

〈**Ans.**〉

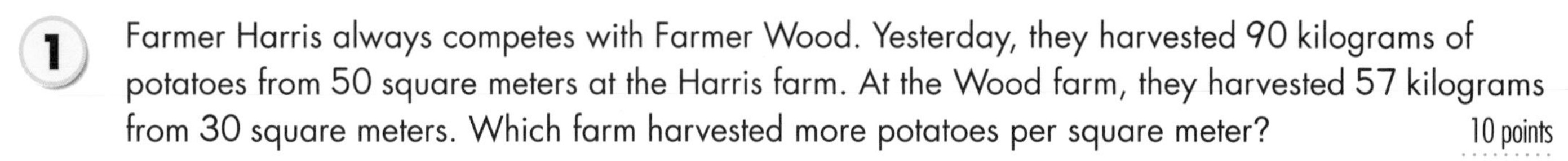

6 Ryan and Kevin both have gardens behind their houses. Ryan planted 360 carrots in 40 square yards behind his house. Kevin planted 460 carrots in 50 square yards behind his house. Which field is more crowded? 10 points

Ryan's carrots / yd^2

Kevin's carrots / yd^2

〈**Ans.**〉

7 After school, we always have the choice of going to the park or staying on school grounds. Today, 40 children went to play in the park, which is 500 square yards. There are 30 children behind the school, on a field that is 300 square yards. Which is more crowded? 10 points

Park children / yd^2

School children / yd^2

〈**Ans.**〉

8 It is time for pep rallies in our town. School A got 820 students together in their 600-square-yard gym. School B had 782 students in their 580-square-yard gym. Which gym was more crowded? 10 points

School A students / yd^2

School B students / yd^2

〈**Ans.**〉

9 One country has an area of 231,000 square kilometers. 93,080,000 people live there. Calculate its population density. Round your answer to the nearest tenth. 10 points

Population (people)		Area (km^2)		Population density ($people/km^2$)
	÷		=	

〈**Ans.**〉

10 My town has an area of 38 square kilometers and a population of 7,824. My aunt's town, just across the bridge, has an area of 42 square kilometers and a population of 9,240. Which town has the higher population density? 10 points

My town people / km^2

My aunt's town people / km^2

〈**Ans.**〉

Time to put a move on! Excellent.

1 Ted bikes every day. Today, he rode 80 miles in 2 hours while biking around the lake. How many miles per hour was he going? 10 points

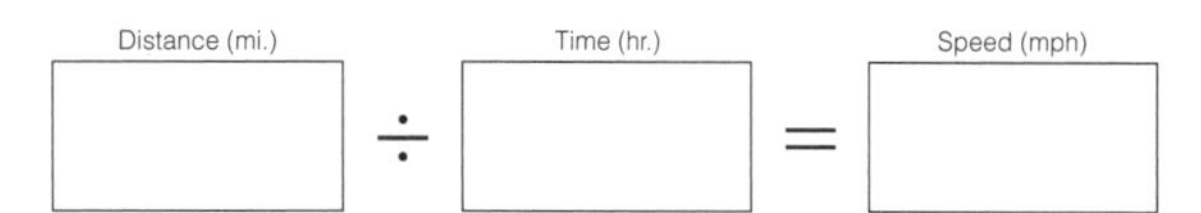

〈Ans.〉 mph

2 I like to go a lot slower than Ted when I bike. Today, I went only 2,700 feet in 15 minutes. How many feet per minute was I going? 10 points

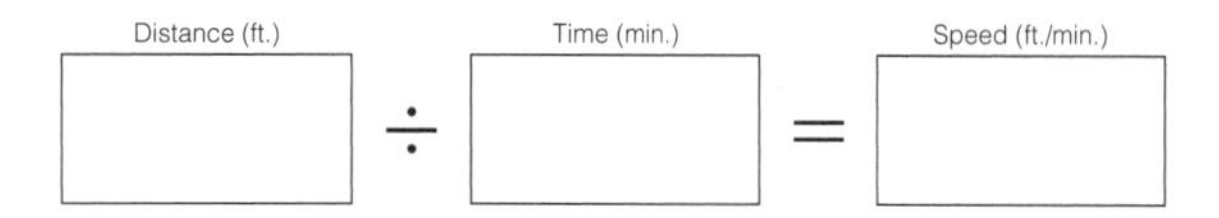

〈Ans.〉 ft./min.

3 Lizzy loves her horse. Today, she let him run as fast as he wanted to for a little bit. She counted that he went 120 meters in 8 seconds. How many meters per second was her horse going? 10 points

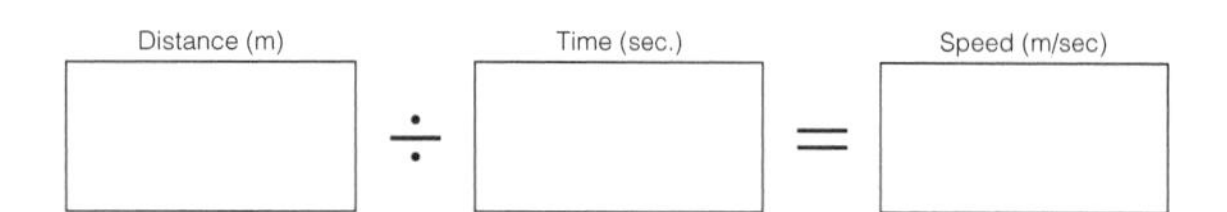

〈Ans.〉 m/sec

4 Kurt is training for his track meet coming up. Today, he ran 4,000 feet in 25 minutes as a warm-up. How many feet per minute was he going? 10 points

〈Ans.〉

5 Brenda's father was late coming home from his business trip. He decided to drive a little faster and went 200 miles in the last 2.5 hours he was driving. How many miles per hour was he going? 10 points

〈Ans.〉

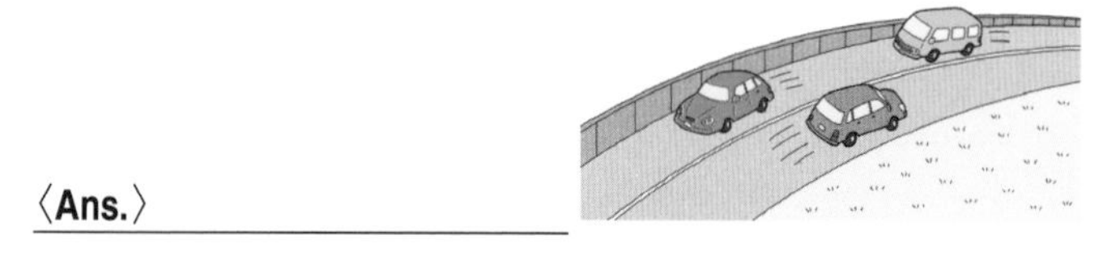

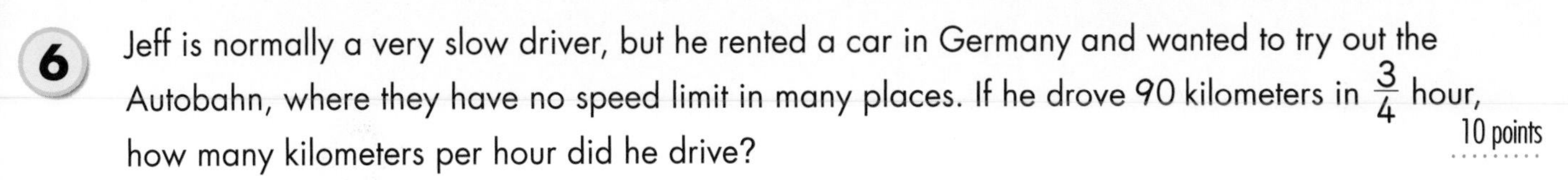

6 Jeff is normally a very slow driver, but he rented a car in Germany and wanted to try out the Autobahn, where they have no speed limit in many places. If he drove 90 kilometers in $\frac{3}{4}$ hour, how many kilometers per hour did he drive? 10 points

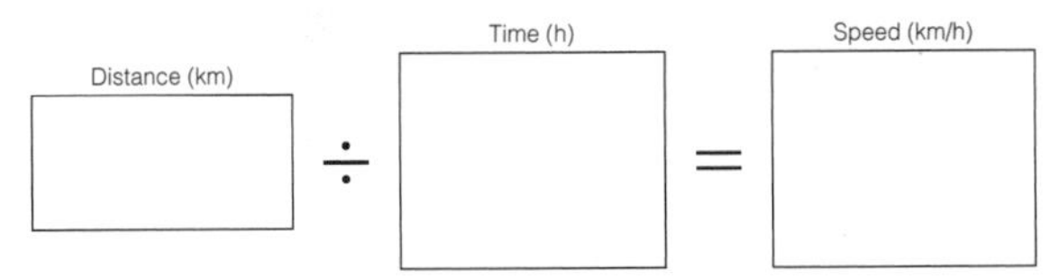

⟨**Ans.**⟩ ____________

7 He decided to try and slow down, but he liked driving fast. If he drove 90 kilometers in the next 45 minutes, how many kilometers per hour was he going? Rewrite the minutes as a fraction of an hour to solve the problem. 10 points

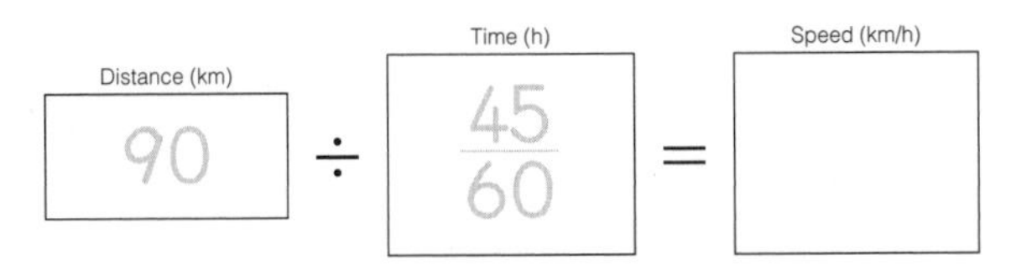

⟨**Ans.**⟩ ____________

8 Zach's house is 24 kilometers away from his uncle's house. If it took him 40 minutes to drive the tractor to his uncle's house, how many kilometers per hour was he going? Rewrite the minutes as a fraction of an hour to solve the problem. 10 points

⟨**Ans.**⟩ ____________

9 Megan wanted to visit her cousin in the next town, but her mother took the car to the store. She biked the 4 kilometers in 20 minutes. How many kilometers per hour was she going? 10 points

⟨**Ans.**⟩ ____________

10 Lance is training to run the 100 meters in the track finals. Today, he ran 150 meters in 24 seconds. How many meters per minute was he running? 10 points

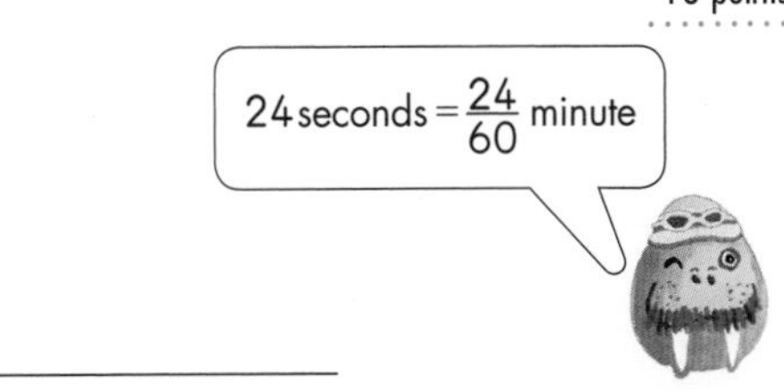

⟨**Ans.**⟩ ____________

How many pages per hour can you go?
Just kidding!

1 Lee's grandfather drives so slow. Today, he's driving 40 miles per hour. How far will he go in 3 hours? 10 points

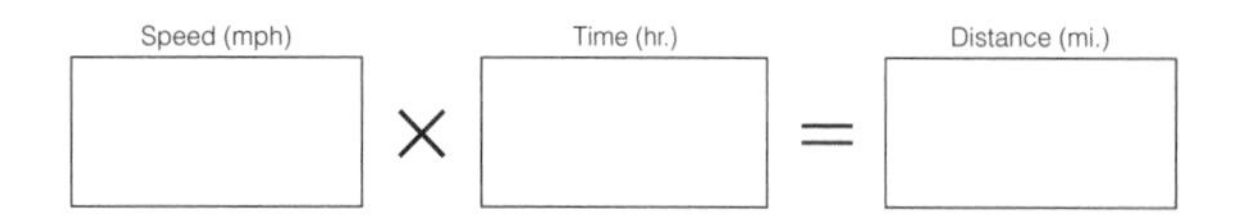

〈Ans.〉

2 My train to school goes 64 kilometers per hour. The train takes 2 hours 30 minutes to go from the first stop to the last stop. How far is it from the first stop to the last stop? 10 points

〈Ans.〉

3 Katie's bike had a flat tire, and it was raining, so she decided to run to her friend's house as fast as she could. She can run 300 yards per minute and it took her 15 minutes. How far away is her friend's house? 10 points

〈Ans.〉

4 George's bus is the slowest bus in town—he's sure of it. He measured, and it goes about 500 yards in a minute. How far will it go in 25 minutes? 10 points

〈Ans.〉

5 A rocket can go as fast as 5 miles per second. In 5 minutes, how far will it fly? 10 points

〈Ans.〉

6 Lisa took the train to her grandfather's house. If the train went 60 miles per hour, and her train ride took 45 minutes, how far did she go? Rewrite the minutes as a fraction of an hour in order to solve the problem. 10 points

Speed (mph) ☐ × Time (hr.) ☐ = Distance (mi.) ☐

45 minutes = $\frac{\square}{60}$ hour

〈Ans.〉

7 Jocelyn took her bike to the store because the car was in the shop. She went 13 miles per hour and it took her 50 minutes. How far away is the store from her house? Rewrite the minutes as a fraction of an hour in order to solve the problem. 10 points

〈Ans.〉

8 Mr. Medina's express train runs 80 kilometers per hour for 1 hour 45 minutes every morning. How far does it go? 10 points

〈Ans.〉

9 Luke's father wouldn't let him get a motorbike because he thought it was too dangerous. Instead, he bought Luke a moped. The moped goes 40 miles per hour at top speed. How far can Luke go in 1 hour 24 minutes if he goes top speed the whole way? 10 points

〈Ans.〉

10 Ari left his house and began walking 3 miles per hour. It took him 25 minutes to get to the train station. How far is his train station from his house? 10 points

〈Ans.〉

You're doing fine. Keep it up!

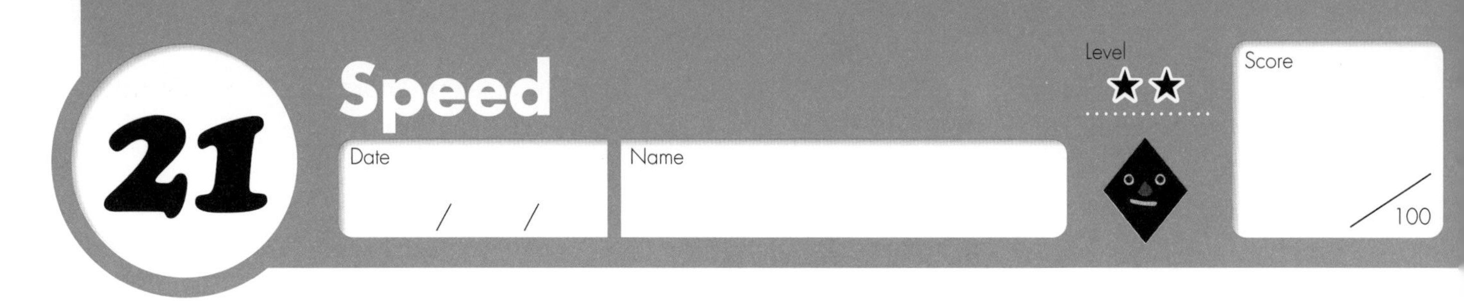

1 Our scout troop has 12 miles to hike today. If we hike 3 miles per hour, how long will it take us?

10 points

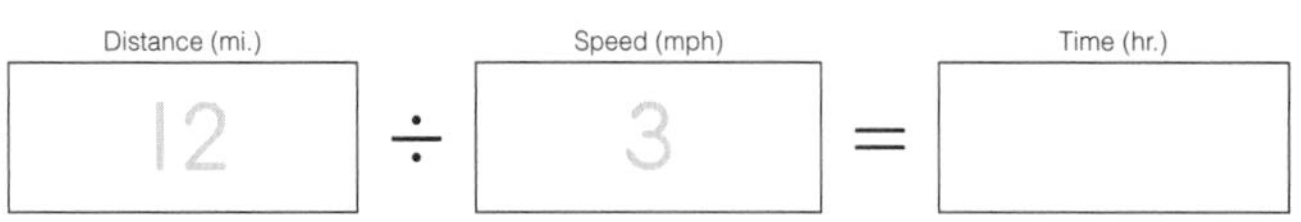

〈**Ans.**〉

2 It is only 910 feet from Hilary's house to the park. If she walks 65 feet per minute, how long will it take her to get to the park?

10 points

〈**Ans.**〉

3 From our shore house to the beach, it is 5 miles. My brother decided to run to the beach and was running 0.1 mile per minute. How long will it take him?

10 points

〈**Ans.**〉

4 Riko got lost in her friend's neighborhood and is driving slowly around the block looking for the right street. If she is driving 10 miles per hour for 15 miles, how long will it take her to find the street she is looking for?

10 points

〈**Ans.**〉

5 Theresa took the train from the West station to the East station. If the distance was 54 miles, and her train ran 0.5 miles per minute, how long did it take her?

10 points

〈**Ans.**〉

6 Billy's train was going 40 kilometers per hour for 34 kilometers. How long did it take? 10 points

$\frac{17}{20}$ hour $= \left(60 \times \frac{17}{20}\right)$ minutes $= 51$ minutes

Distance (km)		Speed (km/hr.)		Time (hr.)
34	÷	40	=	$\frac{17}{20}$

$\frac{17}{20}$ hour = minutes

〈Ans.〉

7 Our plane was going 480 kilometers per hour, and it flew 880 kilometers. How many hours and minutes did it take? 10 points

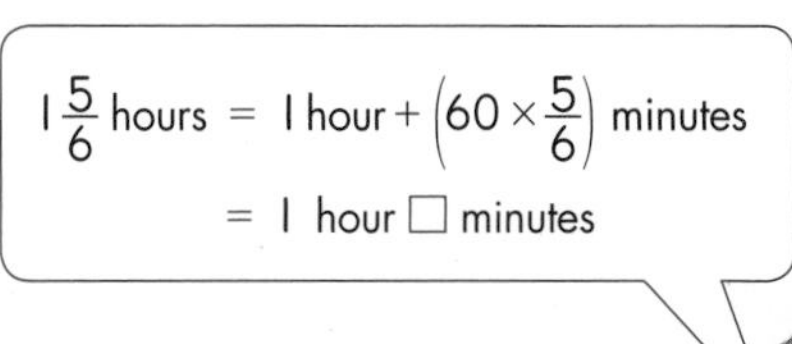

〈Ans.〉

8 Caitlin was slowly biking around the pond for fun. If she was going $\frac{3}{14}$ kilometer per minute for 6 kilometers, how long did it take her? 10 points

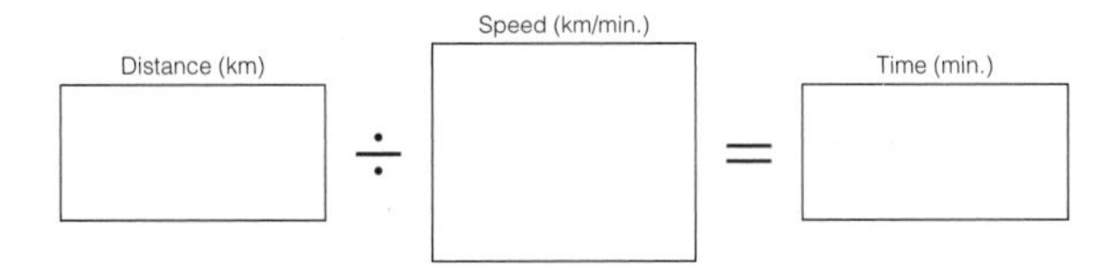

〈Ans.〉

9 Fran walked 3 kilometers per hour through the park on Sunday. If she walked 2 kilometers, how long did it take her? 10 points

〈Ans.〉

10 Louis biked slowly in the rain to the library. If he was going 15 kilometers per hour, and had 11 kilometers to go, how long did it take him? 10 points

〈Ans.〉

How fast can you bike?
Can you figure it out?

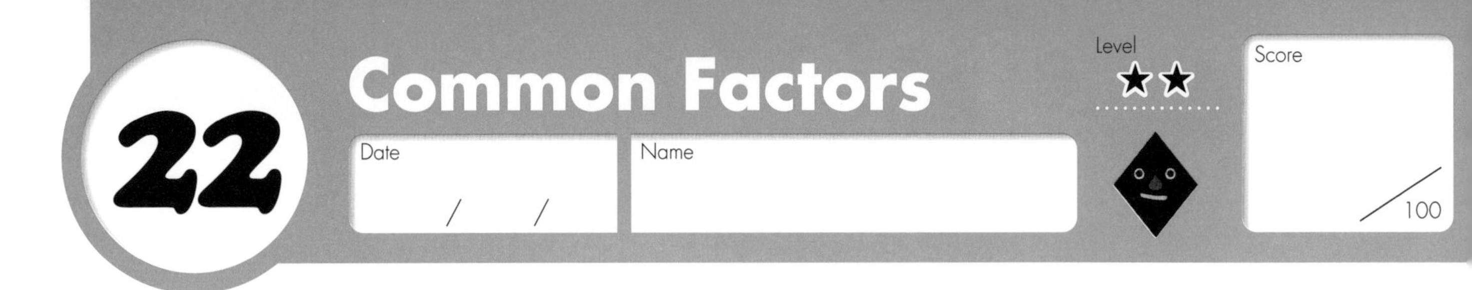

1. Wayne is working in the back room of the bookstore. He's supposed to pile the 4-inch thick encyclopedias in one pile and the 6-inch thick dictionaries in another pile. How tall will the two piles be when they are the same height? 10 points

〈Ans.〉

2. Tom never gets the good jobs on the construction site. Today, he's supposed to pile up the 6-inch wooden boards and the 9-inch blocks next to each other. How tall will the piles be when they are the same height? 10 points

〈Ans.〉

3. The red lights on our holiday decorations blink every 6 seconds. The blue ones blink every 8 seconds. If both lights are plugged in at the same time, how long will it be until they blink together? 10 points

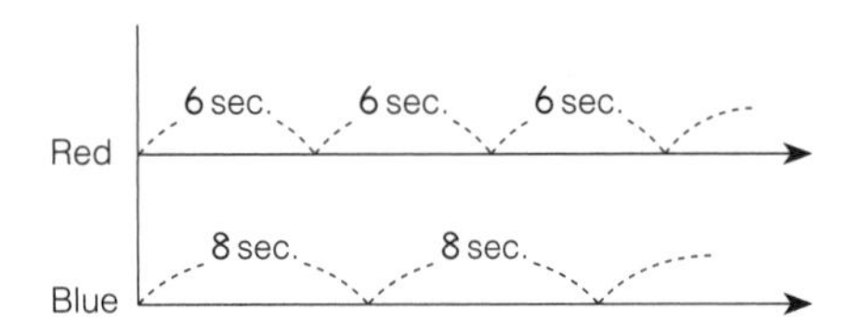

〈Ans.〉

4. At our bus depot, the bus going to West Mountain leaves every 12 minutes. The bus going to East Mountain leaves every 8 minutes. If the first bus in both directions leaves at 7 a.m., what is the next time both buses leave at the same time? 10 points

〈Ans.〉

5. At the local race, the #1 car was slower than the #5 car. The #1 car made it around the course in 18 seconds, while the #5 car took 16 seconds. If they keep going, how long will it be until they meet at the starting line again? 10 points

〈Ans.〉

6 You have a set of rectangular cards that are 6 centimeters long and 9 centimeters wide. What is the smallest square you can arrange with these cards? 5 points per question

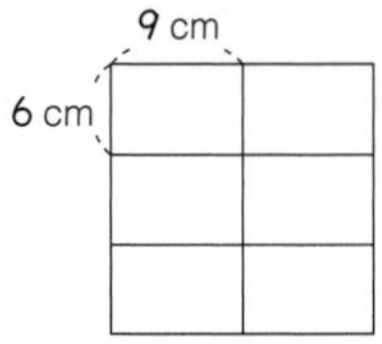

9 cm
6 cm

(1) How long is each side of the square?

⟨Ans.⟩

(2) How many cards do you need?

⟨Ans.⟩

7 Now you have a set of rectangular cards that are 8 centimeters long and 12 centimeters wide. How many cards will you need to make the smallest square possible with these cards? 10 points

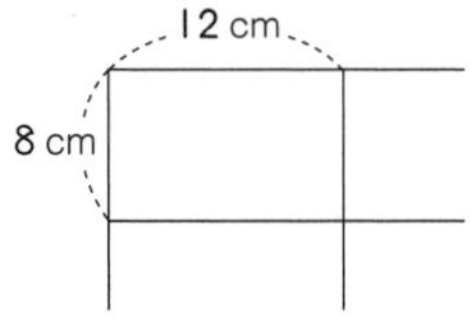

12 cm
8 cm

⟨Ans.⟩

8 Answer the following questions as if you had a set of rectangular cards that are 6 inches long and 8 inches wide. 10 points per question

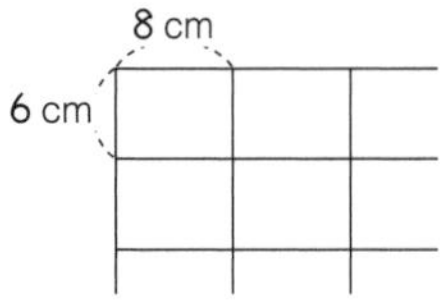

8 cm
6 cm

(1) How many cards would you need to make the smallest square possible with these cards?

⟨Ans.⟩

(2) How many cards would you need to make the second-smallest square possible with these cards?

⟨Ans.⟩

9 Now you have some blocks that are 6 inches long, 8 inches wide, and 3 inches high. You arranged them as shown below in order to make the smallest cube possible. How many blocks did you need to use? 10 points

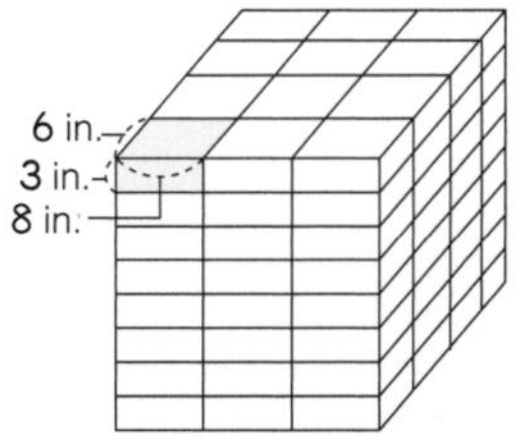

6 in.
3 in.
8 in.

cube : length = width = height

⟨Ans.⟩

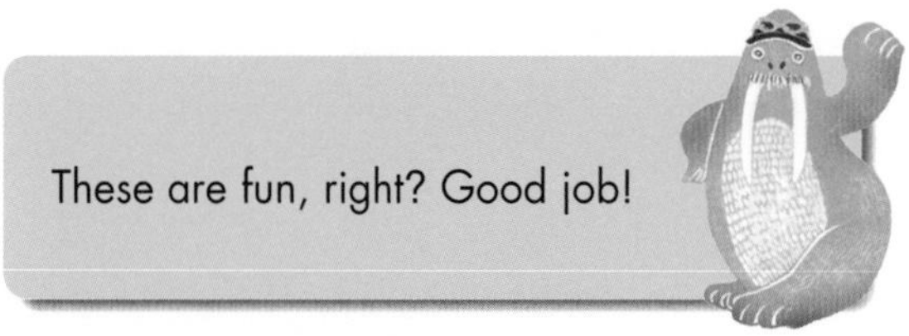

These are fun, right? Good job!

23 Common Factors

Level ★★

Date / /

Name

Score /100

1 Philip has 12 candies he wants to share with his friends. Below are pictures of different situations. If each person gets the number of candies shown in each box below, how many people will get candy? Write the appropriate number in each box.

6 points per box

A per person

12 candies, 6 people

B per person

12 candies, ☐ people

C per person

12 candies, ☐ people

D per person

12 candies, ☐ people

2 You have 8 candies and 12 cookies, and you want to divide them equally among your friends. If you divide them as shown in the pictures below, how many people will get sweets? Write the appropriate number in each box.

6 points per question

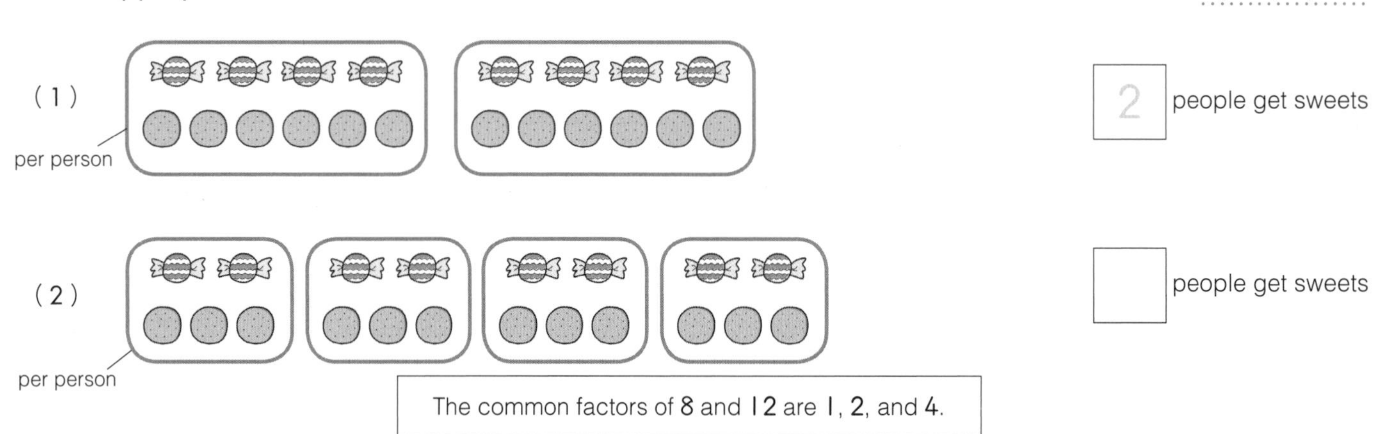

The common factors of 8 and 12 are 1, 2, and 4.

3 If I have 9 candies and 6 cookies, and want to divide them up so that some of my friends get an equal amount of each sweet and there are no remaining sweets, how many people get sweets? (The answer is more than one.)

10 points

〈Ans.〉

4 We have 8 notebooks and 12 pencils in our group today. How many people can get notebooks and pencils if we want to divide each supply equally with no remainder? (Please give all the different combinations that work other than one.) 10 points

⟨**Ans.**⟩

5 Our scout troop has 16 apples and 24 oranges for snack time. We want to divide them up so everyone gets an equal amount of each fruit, and there are no leftovers. How many people can get fruit? (Please give all the different combinations that work other than one.) 10 points

⟨**Ans.**⟩

6 Mr. Henderson is dividing us up into groups for a science project. There are 32 boys and 24 girls in our class. If he wants to make the smallest groups possible with equal amounts of boys and girls in each group and no remainders, how many groups should he make? 10 points

⟨**Ans.**⟩

7 Melissa brought 24 candies and 30 cookies to the shelter to give away. If she wants to make goodie bags with the same amounts of candies and cookies in each for as many children as possible with no remainders, how many of each sweet will each child get? 10 points

⟨**Ans.**⟩

8 In art class today, our teacher had 36 red sheets of paper and 42 blue sheets of paper. If our teacher wanted to give the most students the same amounts of each type of paper with no remainders, how many sheets of blue and red paper did we each get? 10 points

⟨**Ans.**⟩

9 You have a piece of paper that is 18 inches long and 24 inches wide. If you want to cut the rectangle up into square pieces so that no paper is remaining, how long will the sides of your square pieces be? 10 points

24 in.

18 in.

⟨**Ans.**⟩

These are tough. Good job!

24 Speed & Distance

Date / / Name Level ★★★ Score /100

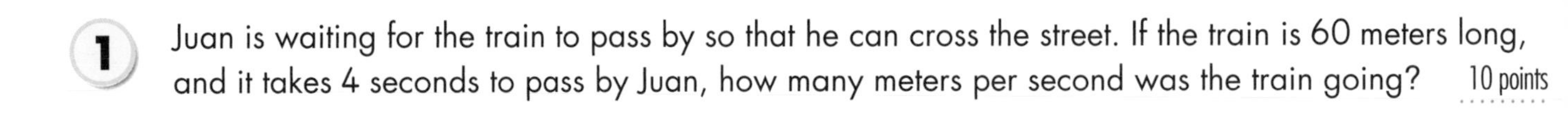

1 Juan is waiting for the train to pass by so that he can cross the street. If the train is 60 meters long, and it takes 4 seconds to pass by Juan, how many meters per second was the train going? 10 points

〈Ans.〉

2 Rob is swimming under the 120-meter long train bridge on the edge of town. A 80-meter long train goes over the bridge, and Rob counts how long it takes to pass completely over the bridge. If it took 5 seconds, how many meters per second was the train going? 10 points

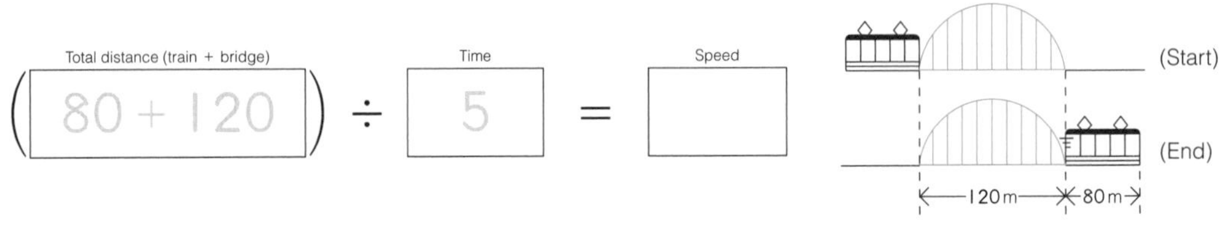

〈Ans.〉

3 Andrew is on a train, and hates tunnels. He counted how long his 80-meter long train took to pass completely through a 240-meter tunnel. If it took 8 seconds, how many meters per second was his train going? 10 points

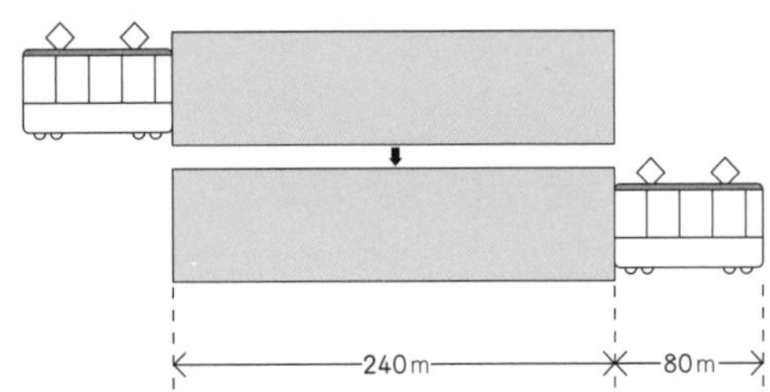

〈Ans.〉

4 If a train is 90 meters long and takes 30 seconds to pass over a 660-meter bridge, how many meters per minute is the train going? 10 points

〈Ans.〉

5 Our train runs 50 meters per second. My little brother loves counting, so he counted how long it took to pass over a 900-meter bridge. If it took 20 seconds, how long is the train? 10 points

Reverse the number sentence: the length of the train is the speed multiplied by the amount of time it took minus the distance of the bridge.

〈Ans.〉

6 If a train going 20 meters per second takes 20 seconds to pass all the way through a 330-meter tunnel, how long is the train? 10 points

⟨Ans.⟩

7 How long would a 120-meter train going 25 meters per second take to go all the way through a 580-meter tunnel? 10 points

⟨Ans.⟩

8 There is a bridge near us, and my little brother always counts how long it takes for a train to go over the bridge. If I know the bridge is 453 meters long, and a 145-meter train is passing over it going 23 meters per second, how many seconds will my brother count? 10 points

⟨Ans.⟩

9 How long will it take a 100-meter train going 1.8 kilometers per minute to pass through a 500-meter tunnel? Answer in seconds. 10 points

⟨Ans.⟩

10 Our train is going through a long tunnel under a mountain near town. If the train is going 86.4 kilometers per hour, and it took us 2 minutes 10 seconds to pass through a 2.97-kilometer tunnel, how long is our train? 10 points

86400 ÷ 60 ÷ 60 = 24
86.4 km/hr. = 24 m/s
2 minutes 10 seconds = 130 seconds

⟨Ans.⟩

I am a big train fan. How about you?

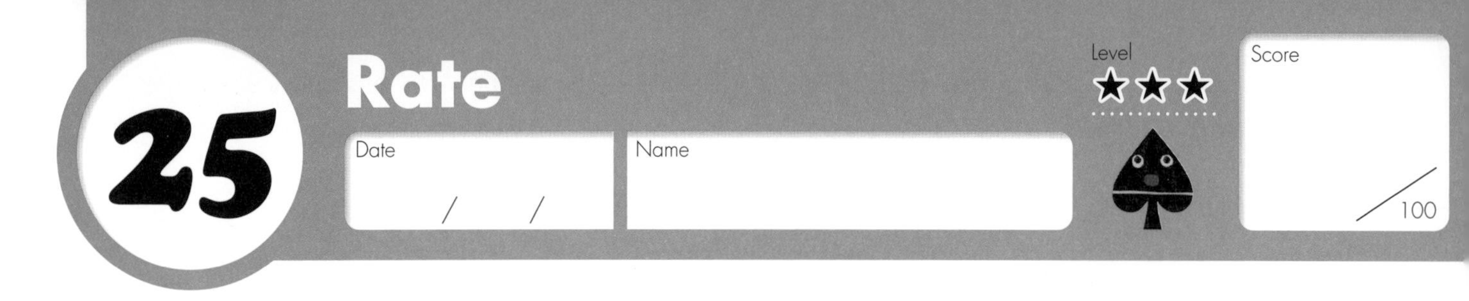

Rate

1. Yugo's house is 5,625 yards away from Gabe's house, and they both decide to leave at the same time. If Yugo walks 60 yards per minute from his house, and Gabe walks 65 yards per minute from his house, how long will it take them to meet? 10 points

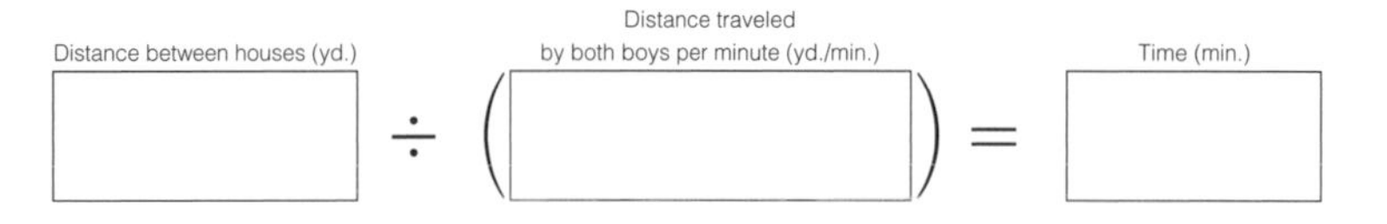

〈Ans.〉

2. Keach and Carmen decide to walk in different directions around a pond. If the distance around the pond is 3,350 yards, and Keach is walking 70 yards per minute, and Carmen is walking 64 yards per minute, when will they meet again? 10 points

〈Ans.〉

3. Tiago thought he was supposed to go to Jason's house, which is 12 kilometers away. Jason thought he was supposed to go to Tiago's house. They both left their houses at the same time. If Tiago is walking 4.2 kilometers per hour, and Jason is walking 4.8 kilometers per hour, how long will it take them to meet? Answer in hours and minutes. 10 points

〈Ans.〉

4. Jonas got a summer job at the local aquarium. Today, he's supposed to fill up a 1,000-liter tank with two hoses. If the water flows from hose A at the rate of 10 liters per minute and from hose B at the rate of 15 liters per minute, how many minutes will it take to fill up the tank? 10 points

〈Ans.〉

5. Ryan and his brother are trying to save up for a game that costs $27. If Ryan saves $1.50 per month, and his brother saves $3 per month, how long will it take for them to save up the correct amount? 10 points

〈Ans.〉

6 Kate and Jeff got into an argument and walked off into separate directions. If Jeff was walking 80 meters per minute, and Kate was walking 70 meters per minute, how long will it take until there is 3 kilometers of distance between them? 10 points

Distance between them (m)		Combined distance traveled per minute (m/min.)		Time (min.)
3,000	÷	(80 + 70)	=	

〈Ans.〉

7 Mike's train station is in the middle of nowhere. The tracks go straight into the distance in each direction. If a train leaves his station going in one direction going 40 kilometers per hour, and another train leaves in the other direction going 50 kilometers per hour, how many hours later will the 2 trains be 225 kilometers apart? 10 points

〈Ans.〉

8 Shannon's brother left to go to the store with his bike, and he was going 125 meters per minute. He forgot his wallet, so Shannon took his brother's wallet and left on his own bike to catch him. If he biked 250 meters per minute, and was 2 kilometers behind his brother, how long will it take him to catch his brother? 10 points

Distance between brothers (m)		Distance caught up per minute (m/min.)		Time (min.)
	÷	( )	=	

〈Ans.〉

9 Jessie and Michelle want to buy a nice tie for their father. Jessie already has $4.50, but Michelle has no money. From now on, Michelle says she will save $3.50 every month, and Jessie will save $2 per month. How long will it take until they have the same amount of money? 10 points

〈Ans.〉

10 Nelson left for his friend's house and was walking 60 meters per minute. 14 minutes later, his mother noticed that he forgot his phone and jumped on to a bike to catch up with him. If his mother is riding at a rate of 200 meters per minute, how long will it take her to catch up to him? 10 points

Distance between them (m)		Distance caught up per minute (m/min.)		Time (min.)
	÷	( )	=	

〈Ans.〉

Okay, let's do something a little different now.

26 Mixed Calculations

Level ★★★

Date / /

Name

Score /100

1 Jordan and his brother got $100 to divide up. If $\frac{1}{3}$ of the money Jordan got was the same as $\frac{1}{2}$ of the money his brother got, how much did each brother get?

10 points per question

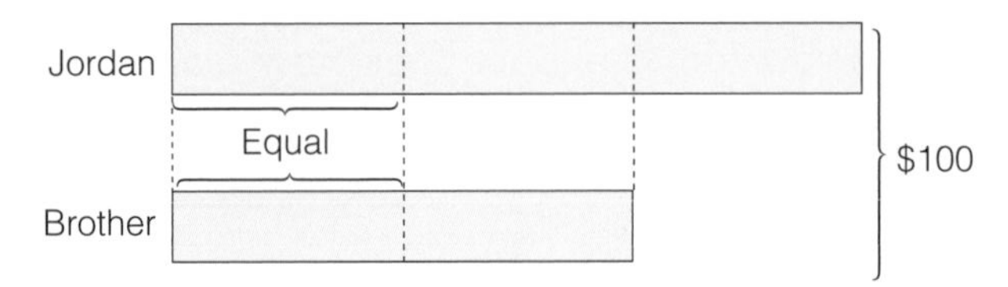

(1) How much money did Jordan's brother get?

$100 \div 5 \times 2$

⟨**Ans.**⟩

(2) How much money did Jordan get?

⟨**Ans.**⟩

2 Rachel and her little sister divided up $120. Rachel reasoned that she was the bigger sister so she should get the bigger share. $\frac{1}{4}$ of the money Rachel got was the same as $\frac{1}{2}$ of the money her sister got. How much did they each get?

20 points

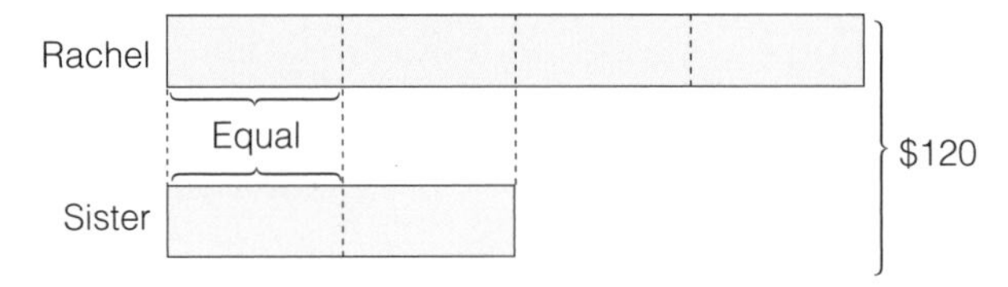

⟨**Ans.**⟩ Rachel Sister

3 Joanne and Veronica divided up 140 stickers. If $\frac{1}{4}$ of the stickers that Joanne got was the same as $\frac{1}{3}$ of the stickers Veronica got, how many stickers did each person get? 20 points

〈**Ans.**〉 Joanne Veronica

4 Rosa and her sister got 88 centimeters of ribbon to share in order to wrap their presents. $\frac{1}{4}$ of the length of ribbon that Rosa got was the same as $\frac{2}{3}$ of the length of ribbon her sister got. How long was Rosa's piece of ribbon? 20 points

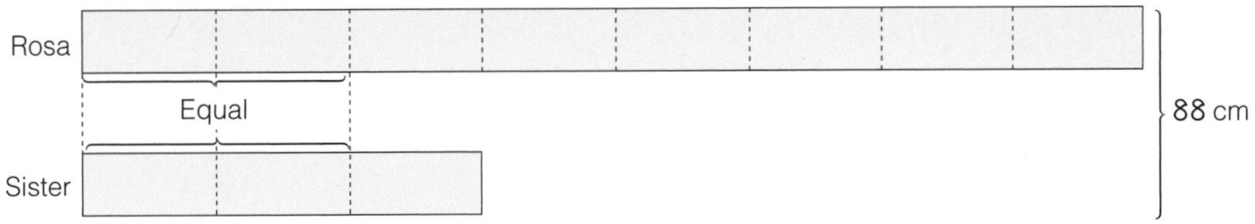

Rosa

Equal

88 cm

Sister

〈**Ans.**〉

5 Luke and his brother were both working on art projects, but there was only 140 inches of tape left. $\frac{1}{2}$ of the length of tape that Luke got was the same as $\frac{3}{4}$ of the length of tape he gave his brother. How long was Luke's piece of tape? 20 points

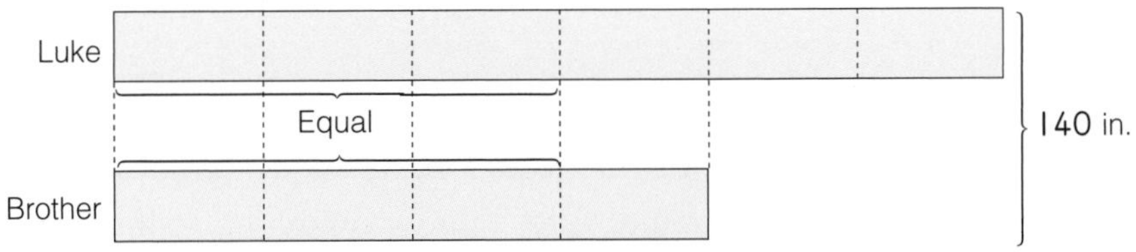

Luke

Equal

140 in.

Brother

〈**Ans.**〉

You're doing really well. Keep it up!

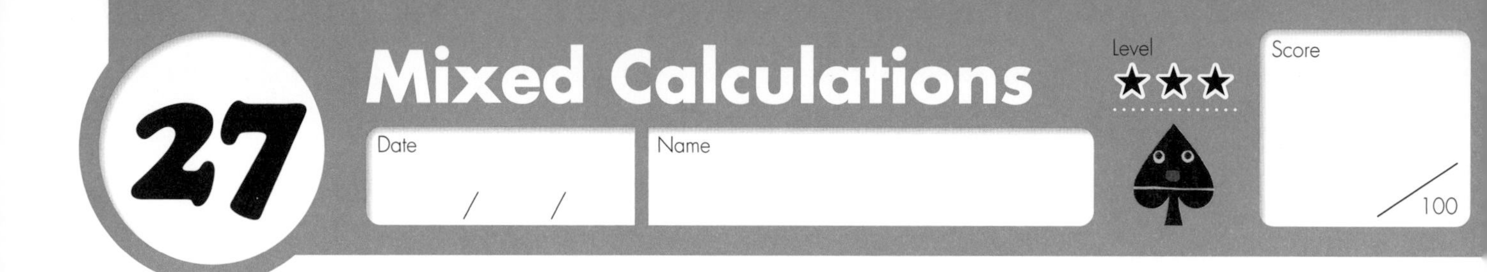

1 The twins, Julian and Damian, both got gumball machines for their birthday. Their mother also gave them 500 gumballs to split. They played a game of basketball to decide who would get more gumballs. Because Julian won, he got twice as many gumballs as Damian, plus 20 more. How many gumballs did they each get?

10 points per question

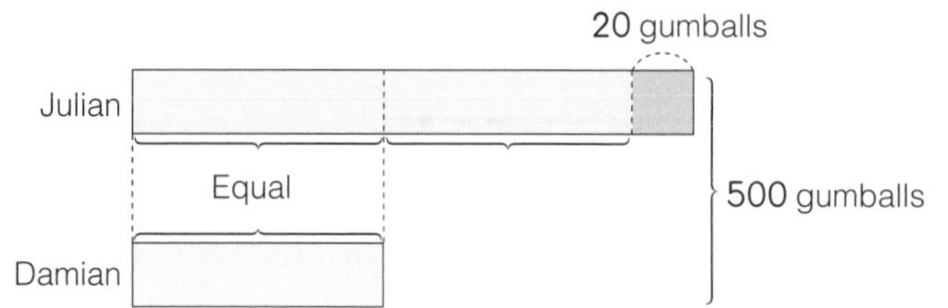

(1) How many gumballs did Damian get?

$(500 - 20) \div 3 =$

〈**Ans.**〉

(2) How many gumballs did Julian get?

〈**Ans.**〉

2 Sean is working at the farm for the summer. He's supposed to divide 140 oranges into a small box and a big box. The number of oranges in the big box is supposed to be 3 times more than the small box, plus 20 more. How many oranges are in each box?

10 points

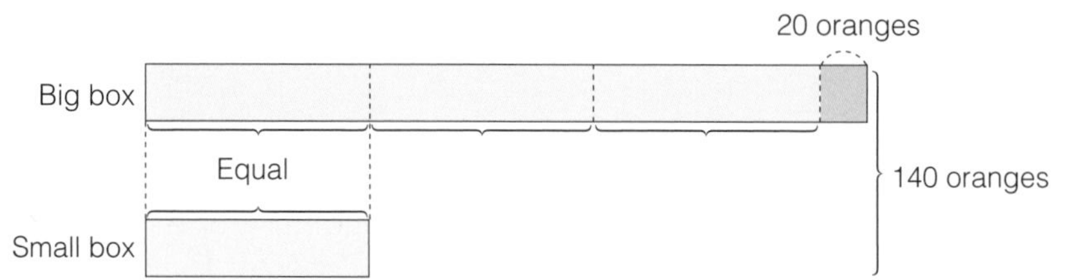

〈**Ans.**〉 Big box Small box

3 I'm making a fruit salad so I bought some apples and pears at the supermarket. 3 of the apples were rotten so I threw them away when I got home. After I threw away the rotten apples, I had 3 times as many apples as pears. If I bought 35 pieces of fruit in all, how many apples do I have?

20 points

〈**Ans.**〉

4 Billy and his little brother split up the $110 their father gave them. Billy got $10 less than twice what is little brother got. How much did they get each?

10 points per question

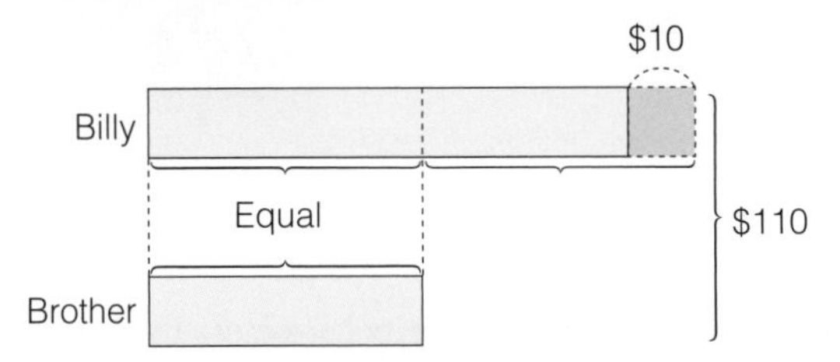

(1) How much did his little brother get?

$(110+10) \div 3=$

〈Ans.〉

(2) How much did Billy get?

〈Ans.〉

5 Kris, Jung and Liz received $250 from Mr. Lawrence. They played a series of rock-paper-scissors games to figure out how much each would get. Kris gets $20 less than twice what Jung will get. Liz gets $30 less than three times what Jung will get. How much did they each get?

15 points

〈Ans.〉 Kris Jung Liz

6 Gary, Audrey and Devin have to split up $62 to buy lunch today. Audrey gets $5 more than twice what Gary gets. Devin gets $3 less than three times what Gary gets. How much do they each get?

15 points

〈Ans.〉 Gary Audrey Devin

Why don't they just share equally?
Some people are so selfish.

Mixed Calculations

Score ／100

Date / /　Name

1 Mr. Rose has a construction team with two workers on it. He's trying to evaluate the two workers, so he's named them *A* and *B*. He knows that it takes *A* 10 days to build a brick wall, and that *B* can do the same job in 15 days. If he asks them to build the wall together, how long will it take?

(1) Think of the whole job as a 1. What ratio of the whole job will each worker do each day? Answer in fractions.

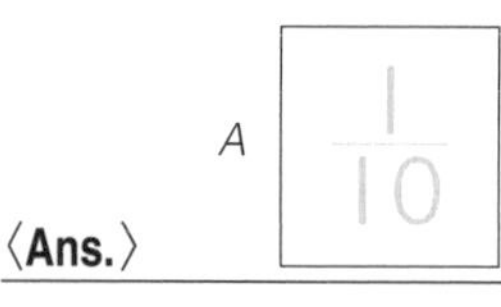

〈Ans.〉 *A* $\frac{1}{10}$ *B* $\frac{1}{15}$

(2) If they work together, what ratio of the work will they do each day?

$\frac{1}{10}+\frac{1}{15}=$

〈Ans.〉

(3) If they work together, how long will it take them to do the job?

$1 \div \frac{1}{6}=$

〈Ans.〉

2 Now, Mr. Rose is thinking about having his two workers paint the house. He knows it would take worker *A* 9 days and worker *B* 18 days to do the job. How long would it take them to paint the house together?

12 points per question

(1) If *A* and *B* work together, what ratio of the whole job will they do each day?

〈Ans.〉

(2) If they work together, how long will it take them to do the job?

〈Ans.〉

3 If Mr. Rose asks *A* to replace the carpeting in the house, it would take him 20 days. *B* would take 30 days. If they work together, how long would it take them to replace the carpet?

10 points

$1 \div \left(\frac{1}{20}+\frac{1}{30}\right)=$

〈Ans.〉

4 Mr. Rose wants to evaluate a new worker on the team, so he's added worker *C* into his calculations. He wants them to retile the bathroom, and he knows that *A* could do it in 10 days, *B* would take 12 days, and *C* would take 15 days. How long will it take all three? 10 points per question

〈**Ans.**〉

5 Now Mr. Rose is used to having *A* and *B* work together, but he wants to know how slow *B* is still. He asked them to replace the appliances in the kitchen, and it took them 8 days. *A* has done the same job in 12 days on another site. How long would it have taken *B* to do the job? 6 points per question

(1) If the whole job is 1, what ratio of 1 did *A* do every day?

〈**Ans.**〉

(2) When *A* and *B* work together, the ratio of work per day is $\frac{1}{8}$. What ratio is *B* doing?

〈**Ans.**〉

(3) How many days would it take *B* to finish on his own?

〈**Ans.**〉

6 *A* and *B* can replace the windows in the bedrooms in 9 days. *A* has done the work before in 12 days. How many days would it take *B* to replace the windows by himself? 10 points per question

〈**Ans.**〉

7 *A* and *B* built the back deck in 15 days. *A* has finished a similar deck in 20 days before. How many days would it take *B* to finish the back deck? 10 points per question

〈**Ans.**〉

Mr. Rose needs some new workers.
You're doing great, though!

29 Mixed Calculations

Level ★★★

Date / /

Name

Score /100

1 Sharon opened up her piggy bank that she found under her bed. It has 20 coins, and they are all nickels and dimes. If the total value is $1.30, how many nickels and dimes does she have?

5 points per question

(1) If all the coins were dimes, how much more money would she have than she actually has?

$10 \times 20 - 130 =$

〈**Ans.**〉

(2) If you then took out 1 dime and replaced it with a nickel, how much less would the total value be?

〈**Ans.**〉

(3) How many dimes would you have to replace with nickels to get to the actual value?

〈**Ans.**〉

(4) How many dimes are there?

〈**Ans.**〉

2 Carol is looking at the turtles and the cranes in the pond at the zoo. Cranes have 2 legs and turtles have 4. There are 10 animals in the pond, and 26 legs. How many cranes and turtles are there?

5 points per question

(1) If all the animals were turtles, how many more legs would be in the pond than are actually there now?

〈**Ans.**〉

(2) How many turtles would you have to replace with cranes in order to get to the correct number of legs?

〈**Ans.**〉

(3) How many turtles are in the pond?

〈**Ans.**〉

3 Mr. Pearman went to buy stamps in order to send all of his holiday cards out. He bought some 50¢ stamps and some 20¢ stamps. If he bought 12 stamps and paid \$4.50, how many of each stamp did he buy?

5 points per question

(1) Start by pretending he bought all 50¢ stamps. Then try to find out how many 50¢ stamps he would have to replace with 20¢ stamps to get the right value.

$(50 \times 12 - 450) \div (50 - 20) =$

⟨Ans.⟩

(2) How many 50¢ stamps did he buy?

⟨Ans.⟩

4 Cheryl's mother is looking at pencils for her daughters. There are some pencils for \$0.80 and some pencils for \$1.00. She bought a dozen mixed pencils and paid \$10.00.

10 points per question

(1) How many pencils did she buy for \$0.80?

⟨Ans.⟩

(2) How many pencils did she buy for \$1.00?

⟨Ans.⟩

5 Mr. Woo is building a fence for his garden with bamboo sticks. Some of his bamboo sticks are 5 feet long and some are 8 feet long. When he connected 10 of them together, he had 65 feet in perimeter.

10 points per question

(1) How many 5-foot sticks did he connect?

⟨Ans.⟩

(2) How many 8-foot sticks did he connect?

⟨Ans.⟩

6 Neil went to the corner store for some gum before his road trip. They had gum for \$0.50 and gum that cost \$0.70. He bought 16 packs and paid \$10. He got \$1 in change. How many \$0.50 packs of gum did he buy?

15 points

⟨Ans.⟩

Wow, this is tough. Way to go!

Mixed Calculations

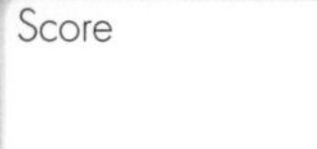

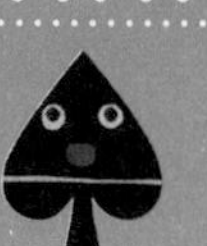
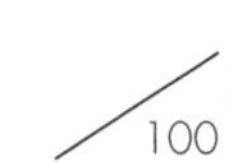

Level ★★★

Date / / Name

Score /100

1 Mr. Chin has some pencils and some students. He needs 12 more pencils in order to be able to give everyone 5 pencils each. If he gives everyone 3 pencils, there are no pencils remaining. How many students and pencils does he have?

5 points per question

(1) How many students are there?

(Hint: If he had 12 more pencils, he could give each child 2 more pencils.)

$12 \div 2 =$

⟨**Ans.**⟩

(2) How many pencils are there?

⟨**Ans.**⟩

2 In art class, Mrs. Harries is handing out paper clips to each person. She needs 15 more clips in order to give everyone 7 clips each. If she gives everyone 4 clips each, there is no remainder. How many paper clips and students does Mrs. Harries have?

5 points per question

(1) How many students are there?

$15 \div (7 - 4) =$

⟨**Ans.**⟩

(2) How many paper clips does she have?

⟨**Ans.**⟩

3 A bunch of trick-or-treaters are at Juliette's door on Halloween. If she gives 7 candies to each person, she'll have no candy left. She needs 18 more candies to give everyone 10 candies each. How many children are at her door, and how many candies does she have?

10 points

⟨**Ans.**⟩

4 Mrs. Osbourne has some grapes to hand out to the children in her kindergarten class. If she gives each child 3 grapes, she'll have 12 grapes left. If she gives each child 5 grapes, she'll have no grapes left. 10 points per question

(1) How many children are in her class?

(Hint: She can give them 2 more grapes each if she uses the 12 remaining grapes.)

$12 \div 2 =$

〈Ans.〉

(2) How many grapes does she have?

〈Ans.〉

5 Mr. Ramirez called for a pop quiz this morning. He gave some paper to our class. If he gives us 6 sheets of paper each, he'll have 16 sheets of paper left. If he gives us each 8 sheets, he'll have no sheets left. 10 points per question

(1) How many students are there in my class?

$16 \div (8 - 6) =$

〈Ans.〉

(2) How many sheets of paper does Mr. Ramirez have?

〈Ans.〉

6 Adam ended up in charge of the children at his family reunion. He decided to give them bouncy balls to play with. If he gave each child 5 balls, he'd have 4 balls remaining. He needs another 6 balls in order to give each child 7 balls. How many children is Adam in charge of? 15 points

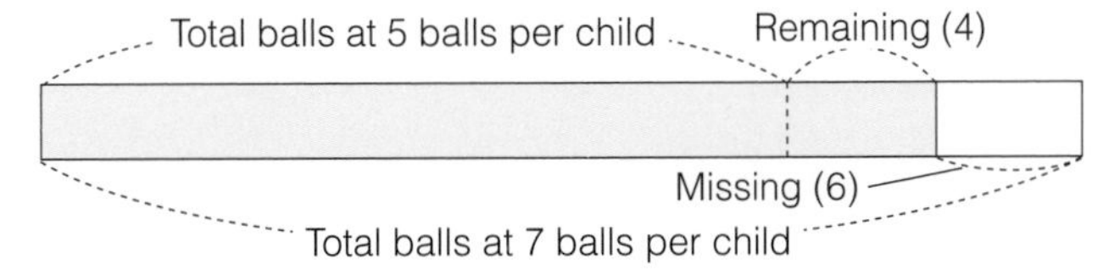

$(4 + 6) \div (7 - 5) =$

〈Ans.〉

7 The arts teacher gave her class some pieces of cardboard for their next project. If she gave her class 5 pieces each, she would have 11 pieces remaining. In order to give them 7 pieces each, she would need 9 more pieces of cardboard. How many students are in her class? 15 points

〈Ans.〉

Phew. Now let's try something totally different!

31 Proportion

Level ★★

Date / /

Name

Score /100

Don't forget!

The table pictured here shows the relationship between time and volume for a sink that is filling up with water. In this example, the sink is filling up at 2 liters per minute.

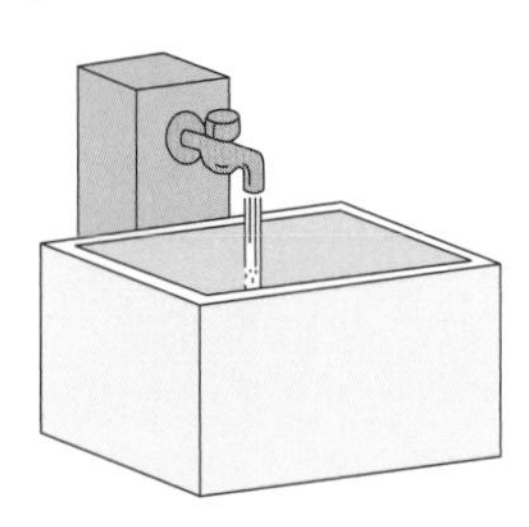

Time (min.)	1	2	3	4	5	6	…
Volume of water (L)	2	4	6	8	10	12	…

(twice, 3 times)

If time is doubled or tripled, then the amount of water in the sink is doubled or tripled. This is because the volume of water is increasing in **proportion** to the time.

1 The tables pictured below show different relationships between time and the depth of water in a sink. Is the depth of water changing in proportion to the time? Put a ✓ next to it if the water is rising in proportion to time, and an × if not.

10 points per question

(1)

Time (min.)	1	2	3	4	5	6
Depth of water (in.)	3	6	9	12	15	18

(twice, 3 times)

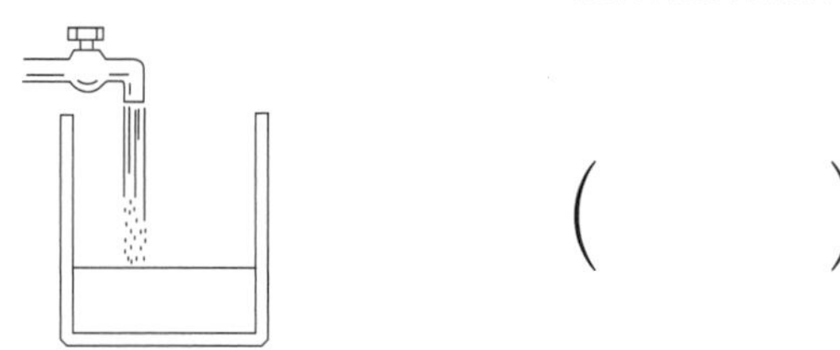

( )

(2)

Time (min.)	1	2	3	4	5	6
Depth of water (in.)	2	3	4	6	10	16

(twice, 3 times / ?, ?)

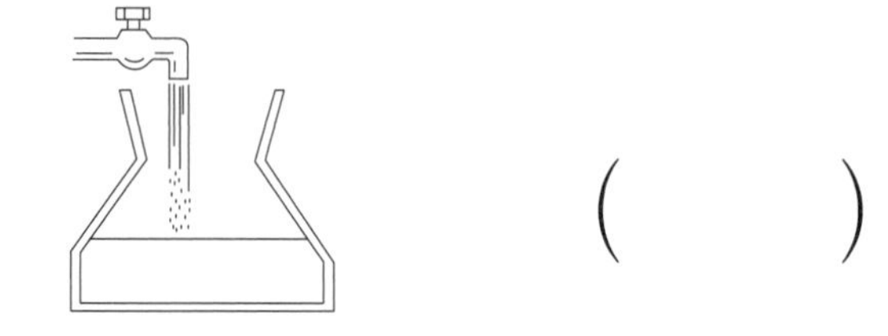

( )

(3)

Time (min.)	1	2	3	4
Depth of water (in.)	5	6	7	8

( )

(4)

Time (min.)	1	2	3	4
Depth of water (in.)	5	10	15	20

( )

(5)

Time (min.)	2	4	6	8
Depth of water (in.)	10	20	30	40

( )

(6)

Time (min.)	2	4	6	8
Depth of water (in.)	10	15	20	30

( )

2 Fill in the following tables appropriately. Put a ✓ next to each relationship that is proportional, and an × if it is not proportional.

8 points per question

(1) The volume of water in this sink is rising at 3 liters per minute.

Time (min.)	1	2	3	4	5	6	7	…
Volume of water (L)	3							…

( )

(2) The relationship between the age of a mother and her child on her child's birthday.

Mother's age	25	26	27	28				…
Child's age	1	2						…

( )

(3) The relationship between number of 70¢ loaves of bread bought and the total cost.

Number of loaves	1	2	3					…
Cost ($)	0.70							…

( )

(4) The relationship between numbers of sides in a polygon and sum of its interior angles.

Number of sides	3	4	5	6	7	8	9	…
Sum of interior angles (°)	180	360						…

( )

(5) The relationship between time and distance traveled for a train going 80 kilometers per hour.

Time (hrs.)	1	2	3	4				…
Distance (km)	80							…

( )

How are you doing with this? Good!

32 Proportion

Level ★★

Date / / Name

Score /100

1 The table pictured here shows the relationship between time and volume of water in a certain sink.

5 points per question

Time (min.)	1	2	3	4	5	6	7	8	9	10	…
Volume (L)	3	6	9	12	15	18	21	24	27	30	…

(A: from 1 to 3; B: from 4 to 8)

(1) What fraction represents the relationship between the time at the beginning of **A** to the end of **A**?

( )

(2) What fraction represents the relationship between the volume at the beginning of **A** to the end of **A**?

( )

(3) What fraction represents the relationship between the time at the beginning of **B** to the end of **B**?

( )

(4) What fraction represents the relationship between the volume at the beginning of **B** to the end of **B**?

( )

2 The table pictured below shows the relationship between time and the depth of water in a certain sink.

15 points per question

Time (min.)	1	2	3	4	5	6	7	8	9	…
Depth (m)	4	8	12	16	20	24	28	32	36	…

(A: from 2 to 5; B: from 6 to 9)

(1) What fraction represents the relationship between the time at the beginning of **A** to the end of **A**? What is the same relationship for the depth of the water?

(Time) ( ) (Depth) ( )

(2) What fraction represents the relationship between the time at the beginning of **B** to the end of **B**? What is the same relationship for the depth of the water?

(Time) ( ) (Depth) ( )

Don't forget!

Time and volume of water in a sink filling with water

Time X (min.)	1	2	3	4	5	6	…
Volume Y (L)	4	8	12	16	20	24	…

When the volume of water (Y) is proportional to the time (X) as shown in this table, the relationship between Y and X is

$Y = 4 \times X$ (This means that in this case, Y is always 4 times X)

3 Y is proportional to X in each case below. Write the appropriate numbers in each box.

5 points per question

(1)

X	1	2	3	4	5	6	7	…
Y	2	4	6	8	10	12	14	…

($Y =$ [2] $\times X$)

(2)

X	1	2	3	4	5	6	7	…
Y	30	60	90	120	150	180	210	…

($Y =$ [ ] $\times X$)

(3) The relationship between time (X) and volume (Y) in this filling sink is 6 liters per minute.

($Y =$ [ ] $\times X$)

(4) The relationship between time (X) and the distance traveled (Y) by a train going 40 kilometers per hour.

($Y =$ [ ] $\times X$)

4 If Y is proportional to X in the cases shown below, write a number sentence that shows the relationship of Y and X. If Y is not in proportion to X, just write an $\times$ next to the question.

10 points per question

(1) The relationship between the length of a ribbon (X) and the total cost (Y) when the ribbon costs $2 per meter.

(2) The relationship between the volume of water used (X) and water remaining (Y) in a kettle that holds 5 liters.

(3) The relationship between the length of a side (X) and the perimeter of a square (Y).

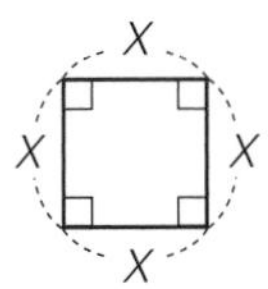

This is really tough. Good job!

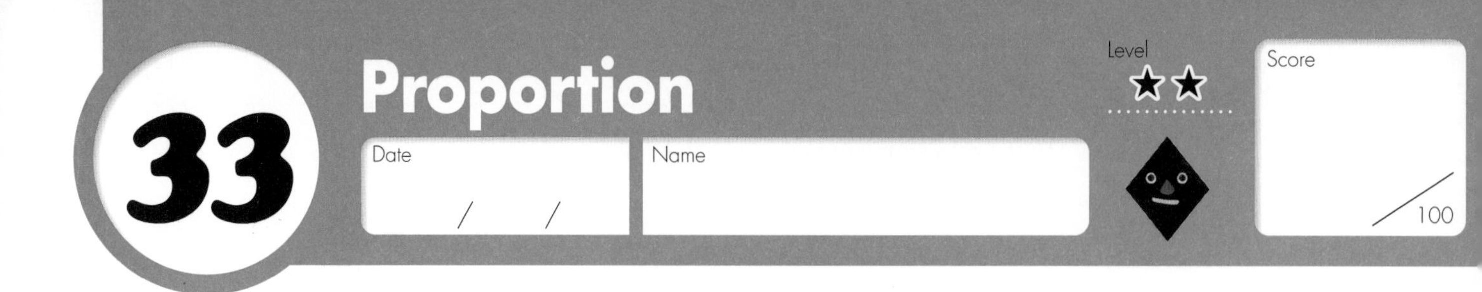

1 The table pictured here shows the relationship between time and the volume of water as it filling a sink. Use the table to draw the graph below.

25 points for completion

Time and volume of water

Time X (min.)	0	1	2	3	4	5	6	7	8	9
Volume Y (L)	0	5	10	15	20	25	30	35	40	45

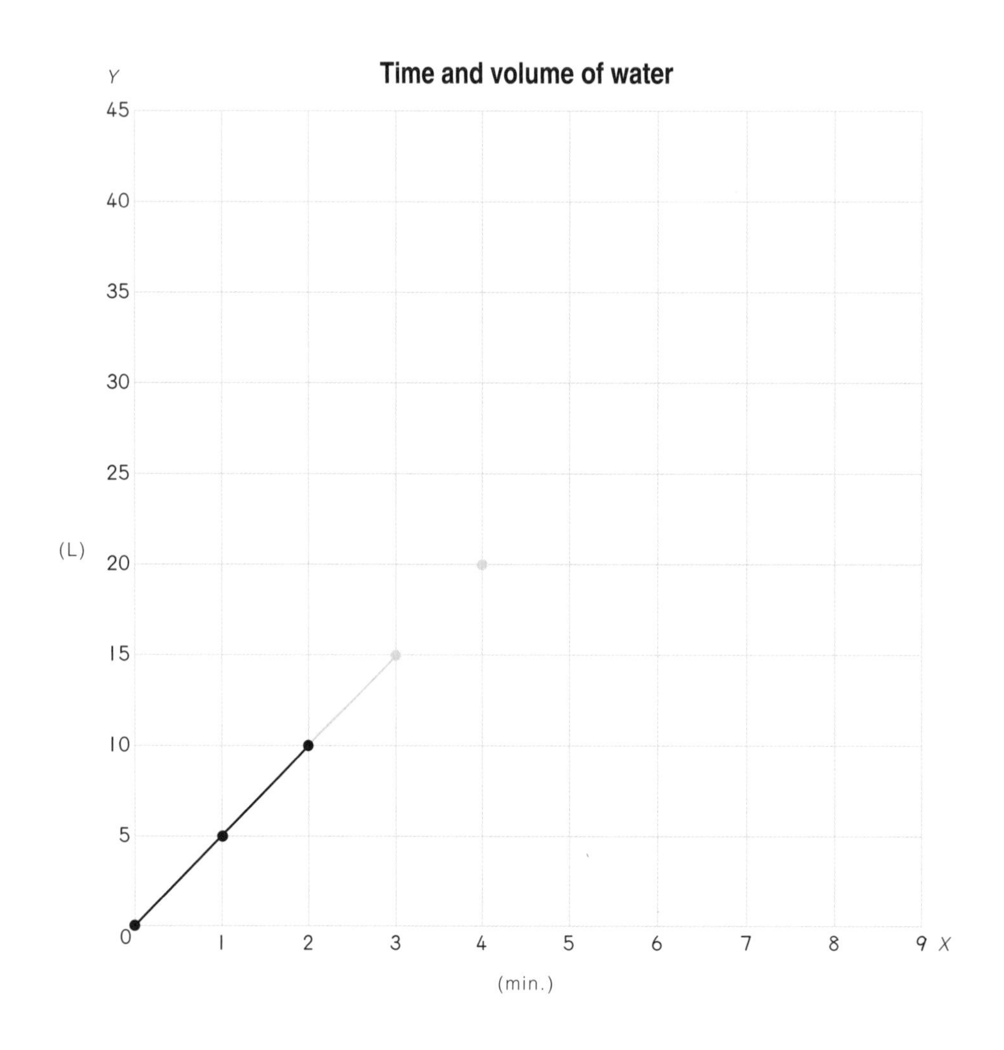

Don't forget!

A diagonal line passing through the point 0 is called a proportion graph.

2 The table below shows that bicycle A is traveling at 0.2 kilometers per minute. Draw the graph below based on the table pictured here. 25 points

Time and distance of bicycle A

Time X (min.)	0	1	2	3	4	5	6	7	8	9	10
Distance Y (km)	0	0.2	0.4	0.6	0.8	1	1.2	1.4	1.6	1.8	2

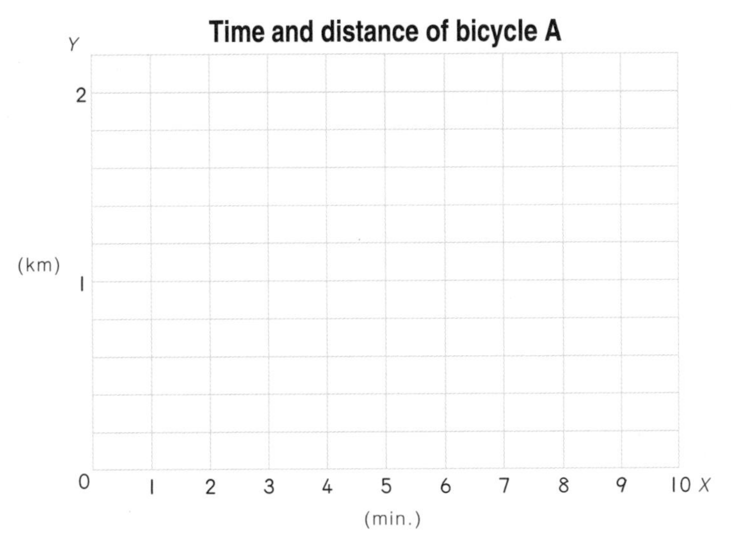

3 The purple ribbon at the craft store costs \$3 per meter. Complete the table based on this proportion, and then draw the graph below. 50 points for completion

Length and cost of purple ribbon

Length X (m)	0	1	2	3	4	5
Cost Y (\$)	0					

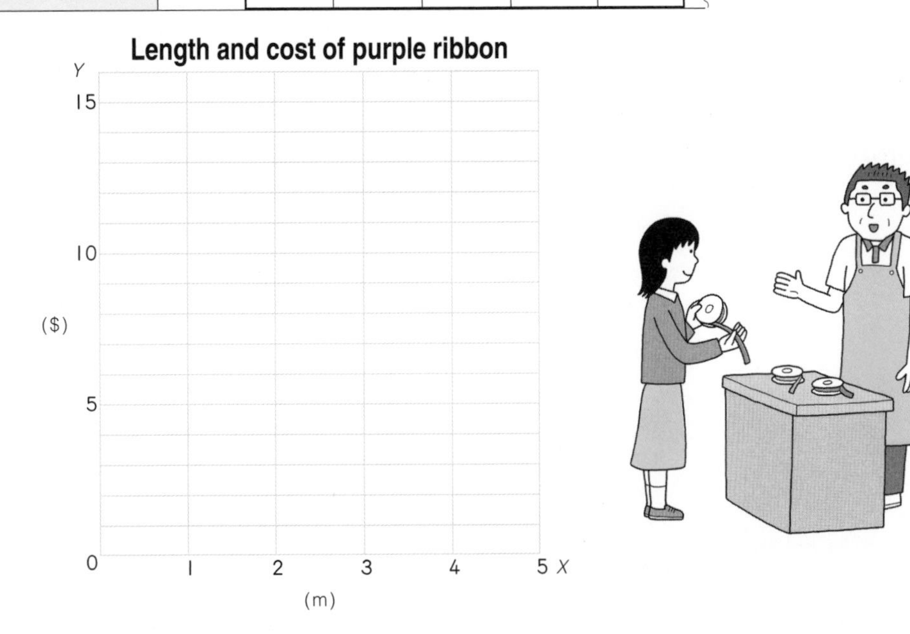

Just a little bit more. You can do it!

Proportion

Date / /

Name

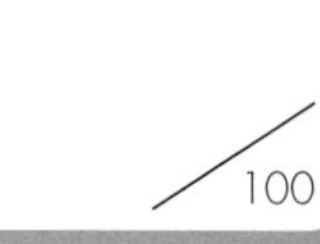

Score

/100

1 The graph to the right shows the relationship between time and the depth of the water filling a sink. Use the graph to answer the questions below. 5 points per question

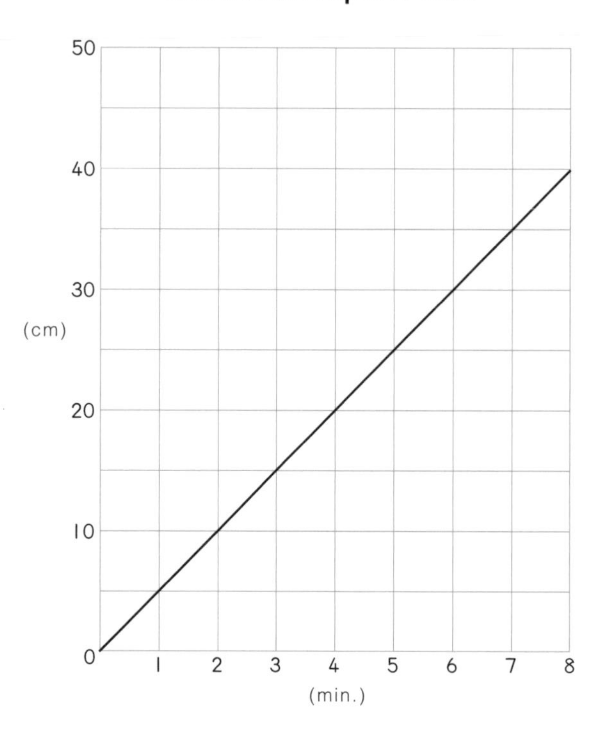

(1) Is the depth of water rising proportionally?

( )

(2) How deep is the water after 6 minutes?

( )

(3) How long has the water been running if the water is 20 centimeters deep?

( )

(4) The sink is filling at the rate of how many centimeters per minute?

( )

2 The graph to the right shows the relationship between the length and weight of a wire. Use the graph to answer the questions below. 5 points per question

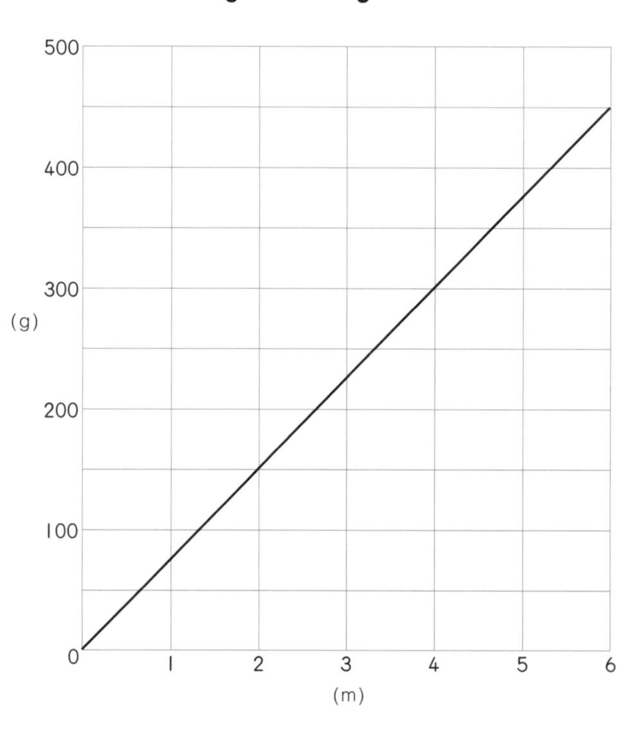

(1) Is the weight of the wire proportional to its length?

( )

(2) How much does the wire weigh when it is 4 meters long?

( )

(3) How many meters long is the wire when it weighs 150 grams?

( )

(4) How many grams per meter does this wire weigh?

( )

3 The graph on the right shows the relationship between the length of a fabric (X) and its cost (Y). Use the graph to answer the questions below. 10 points per question

(1) Complete this table by using values from the graph on the right.

Length and cost of a fabric

Length X (m)	1	2				
Cost Y ($)	1.20					

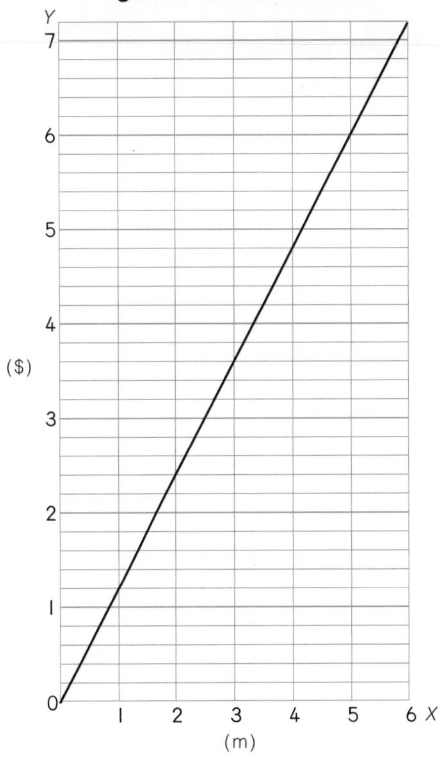

(2) What is $Y \div X$?

( )

(3) Write a number sentence that describes how to calculate Y.

( )

4 The graph on the right shows the relationship between time and the distance a car has traveled. Use the graph to answer the questions below. 10 points per question

(1) Complete this table by using values from the graph on the right.

Time and distance of car A

Time X (hr.)	1	2				
Distance Y (km)						

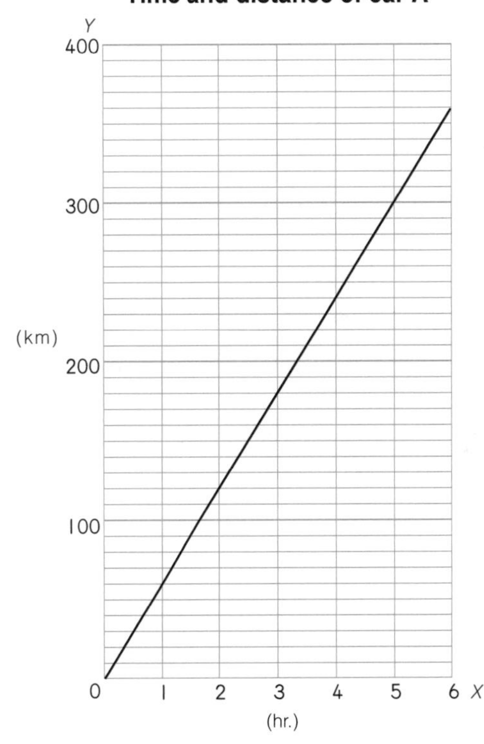

(2) What is $Y \div X$?

( )

(3) Write a number sentence that describes how to calculate Y.

( )

Nice! You did a great job. Now let's review what you've learned.

35 Review

Level ★★★

Date / /

Name

Score /100

1 You have some rectangular cards that are 3 inches long and 5 inches wide. How long is the side of the smallest square you can make with these cards? 10 points

〈**Ans.**〉

2 Charles is trying to cook dinner for the whole family and can only find two containers of almost-empty soy sauce. If he has $\frac{3}{5}$ liter in one bottle, and $\frac{1}{6}$ liter in another bottle, how much soy sauce does he have in all? 10 points

〈**Ans.**〉

3 Maria is still weighing her eggs to figure out what eggs normally weigh. She weighed 5 eggs and got the results shown below. In this group, what does the average egg weigh? 10 points

〔63g 56g 60g 58g 68g〕

〈**Ans.**〉

4 Dan's father keeps track of the mileage on his truck all the time. For example, he knows that he went 420 miles with the last 35 gallons of gas. How many miles per gallon of gas does his truck get? 10 points

〈**Ans.**〉

5 Our train is running 150 miles per hour. If we stay on the train for 3 hours, how far will we go? 10 points

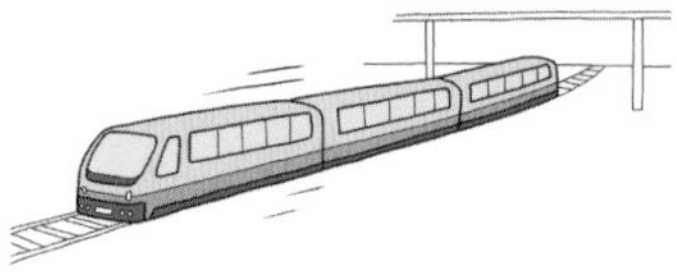

〈**Ans.**〉

6 The recipe calls for $\frac{3}{5}$ liter of milk to make this cake. How much milk will we need to make 4 of these cakes? 10 points

⟨**Ans.**⟩

7 Jim has some paint that is very expensive because it doesn't cover much wall. He can paint $\frac{3}{5}$ square meter with $\frac{1}{3}$ liter of the paint. How much could he cover with 1 liter of paint? 10 points

⟨**Ans.**⟩

8 Fill in the following tables appropriately. The volume of water in this sink is rising at 4 liters per minute. 10 points for completion

Time (min.)	1	2	3	4	5	6	7	…
Volume (L)								…

9 Julia bought $\frac{2}{5}$ kilogram of oranges at the grocery store. That was $\frac{1}{3}$ as many oranges as Tom got. How many kilograms of oranges did Tom buy? 10 points

⟨**Ans.**⟩

10 In science class, Tobin tested some salt water for salt content. He found that there was 5 grams of salt per liter of water. How much salt is in $\frac{7}{10}$ liter of this salt water? 10 points

⟨**Ans.**⟩

Almost there! Way to go!

36 Review

Level ★★★

Date / /

Name

Score /100

1 I have 8 snacks and 12 juice boxes to divide evenly among the people in my study group. How many people can I give food to if I divide it up equally and there are no left overs? (List all the possible combinations other than 1.) 10 points

〈**Ans.**〉

2 Mrs. Harper had $\frac{3}{5}$ liter of olive oil yesterday, but she used $\frac{1}{7}$ liter of it today. How much oil does she have left? 10 points

〈**Ans.**〉

3 The Josephs make 1.5 pounds of garbage per day. How much garbage do they make every 30 days? 10 points

〈**Ans.**〉

4 How many miles per minute is a train going if it goes 162 miles in 3 hours? 10 points

〈**Ans.**〉

5 If a train going is going 2 kilometers per minute, how long will it take it to go 48 kilometers? 10 points

〈**Ans.**〉

6 Bob and Loren have $\frac{3}{4}$ liter of milk. If they divide it equally, how much milk will they each get?
10 points

⟨**Ans.**⟩

7 Mary is trying to grow strong bones. She drank $\frac{3}{5}$ liter of milk yesterday and $\frac{2}{3}$ liter of milk today. What is the ratio of the milk she drank today to the milk she drank yesterday?
10 points

⟨**Ans.**⟩

8 Anne and Betty divided up 180 stamps. If $\frac{1}{5}$ of the stamps that Anne got was the same as $\frac{1}{4}$ of the stamps Betty got, how many stamps did each person get?
10 points

⟨**Ans.**⟩ Anne Betty

9 10 nails in Mr. Carmichael's toolbox weigh 30 grams. How much would 70 of those nails weigh?
10 points

⟨**Ans.**⟩

10 In my state, the population is 1,600,000 and we have 6,500 square kilometers of area. What is the population density in my state? (Round your answer to the nearest tens place.)
10 points

⟨**Ans.**⟩

You did it. Congratulations!

Answer Key Grade 6 Word Problems

1 Review

pp 2, 3

1. $200 \times 1.5 = 300$ — **Ans.** \$300
2. $3.8 \div 0.5 = 7 \text{ R } 0.3$ — **Ans.** 7 pieces, 0.3 m remains
3. $\frac{5}{7} - \frac{1}{7} = \frac{4}{7}$ — **Ans.** $\frac{4}{7}$ m
4. $\frac{4}{5} + \frac{1}{5} = 1$ — **Ans.** 1 L
5. $9 \div 8 = \frac{9}{8}$ — **Ans.** $\frac{9}{8}$ $\left(1\frac{1}{8}\right)$ times
6. $162 \div 1.2 = 135$ — **Ans.** 135 cm
7. $63{,}000 + 58{,}000 = 121{,}000$ — **Ans.** 121,000 people
8. $14 \div 35 = 0.4$ — **Ans.** 0.4
9. $50 \div (50 + 200) = 0.2$ — **Ans.** 20 %
10.

Zone	Number of people	percentage
A	24	48
B	16	32
C	6	12
D	4	8
Total	50	100

2 Review

pp 4, 5

1. $8 \times 8.5 = 68$ — **Ans.** 68 mi.
2. $12 \div 1.5 = 8$ — **Ans.** \$8
3. $32.8 \times 30 = 984$ — **Ans.** 984 g
4. $\frac{7}{9} - \frac{2}{9} = \frac{5}{9}$ — **Ans.** $\frac{5}{9}$ lb.
5. $14.4 \div 4 = 3.6$ — **Ans.** 3.6 yd.
6. $10 \div 7 = \frac{10}{7}$ — **Ans.** $\frac{10}{7}$ $\left(1\frac{3}{7}\right)$
7. $280 \div 350 = 0.8$ — **Ans.** 80 %
8. $20 \times 0.8 = 16$ — **Ans.** 16 lb.
9. $(30 - 27) \div 30 = 0.1$ — **Ans.** 10 %
10. $5 \times 1.2 = 6$ — **Ans.** \$6

3 Fractions

pp 6, 7

1. $\frac{1}{2} + \frac{1}{3} = \frac{5}{6}$ — **Ans.** $\frac{5}{6}$ L
2. $\frac{2}{5} + \frac{1}{4} = \frac{13}{20}$ — **Ans.** $\frac{13}{20}$ ft.^{2}
3. $\frac{5}{12} + \frac{2}{9} = \frac{23}{36}$ — **Ans.** $\frac{23}{36}$
4. $\frac{3}{8} + \frac{3}{5} = \frac{39}{40}$ — **Ans.** $\frac{39}{40}$ m
5. $\frac{1}{4} + \frac{3}{5} = \frac{17}{20}$ — **Ans.** $\frac{17}{20}$ lb.
6. $\frac{3}{8} + \frac{2}{5} = \frac{31}{40}$ — **Ans.** $\frac{31}{40}$ L
7. $\frac{5}{6} + \frac{3}{4} = \frac{19}{12}$ — **Ans.** $\frac{19}{12}$ lb. $\left(1\frac{7}{12}\text{ lb.}\right)$
8. $\frac{1}{2} + \frac{2}{3} = \frac{7}{6}$ — **Ans.** $\frac{7}{6}$ km $\left(1\frac{1}{6}\text{ km}\right)$
9. $\frac{4}{5} + \frac{1}{2} = \frac{13}{10}$ — **Ans.** $\frac{13}{10}$ pt. $\left(1\frac{3}{10}\text{ pt.}\right)$
10. $\frac{3}{4} + \frac{2}{5} = \frac{23}{20}$ — **Ans.** $\frac{23}{20}$ m $\left(1\frac{3}{20}\text{ m}\right)$

4 Fractions

pp 8, 9

1. $\frac{2}{3} - \frac{1}{4} = \frac{5}{12}$ — **Ans.** $\frac{5}{12}$ L
2. $\frac{2}{3} - \frac{1}{2} = \frac{1}{6}$ — **Ans.** $\frac{1}{6}$ L
3. $\frac{4}{5} - \frac{2}{3} = \frac{2}{15}$ — **Ans.** $\frac{2}{15}$ people more
4. $\frac{3}{5} - \frac{2}{7} = \frac{11}{35}$ — **Ans.** He drank $\frac{11}{35}$ liter more today.
5. $\frac{4}{5} - \frac{2}{3} = \frac{2}{15}$ — **Ans.** The corner store is $\frac{2}{15}$ mile further away.
6. $\frac{8}{9} - \frac{1}{3} = \frac{5}{9}$ — **Ans.** $\frac{5}{9}$ L
7. $2 - \frac{3}{4} = \frac{5}{4}$ — **Ans.** $\frac{5}{4}$ lb. $\left(1\frac{1}{4}\text{ lb.}\right)$
8. $\frac{1}{2} - \frac{1}{3} = \frac{1}{6}$ — **Ans.** $\frac{1}{6}$ L
9. $\frac{7}{10} - \frac{2}{5} = \frac{3}{10}$ — **Ans.** $\frac{3}{10}$ mi.
10. $1 - \left(\frac{2}{5} + \frac{1}{4}\right) = \frac{7}{20}$ — **Ans.** $\frac{7}{20}$

5 Fractions

pp 10, 11

1. $3 \times 5 = 15$ — **Ans.** 15 kg
2. $\frac{3}{4} \times 5 = \frac{15}{4}$ — **Ans.** $\frac{15}{4}$ kg $\left(3\frac{3}{4}\text{ kg}\right)$
3. $\frac{1}{6} \times 7 = \frac{7}{6}$ — **Ans.** $\frac{7}{6}$ yd. $\left(1\frac{1}{6}\text{ yd.}\right)$
4. $\frac{3}{5} \times 6 = \frac{18}{5}$ — **Ans.** $\frac{18}{5}$ kg $\left(3\frac{3}{5}\text{ kg}\right)$
5. $\frac{2}{5} \times 3 = \frac{6}{5}$ — **Ans.** $\frac{6}{5}$ L $\left(1\frac{1}{5}\text{ L}\right)$
6. $\frac{2}{3} \times 5 = \frac{10}{3}$ — **Ans.** $\frac{10}{3}$ m^2 $\left(3\frac{1}{3}\text{ m}^2\right)$
7. $\frac{3}{4} \times 6 = \frac{9}{2}$ — **Ans.** $\frac{9}{2}$ L $\left(4\frac{1}{2}\text{ L}\right)$
8. $\frac{3}{4} \times 4 = 3$ — **Ans.** 3 yd.
9. $\frac{5}{6} \times 12 = 10$ — **Ans.** 10 m
10. $\frac{4}{5} \times 10 = 8$ — **Ans.** 8 kg

6 Fractions
pp 12, 13

1. $12 \div 3 = 4$ — Ans. 4 lb.
2. $\frac{7}{10} \div 3 = \frac{7}{30}$ — Ans. $\frac{7}{30}$ lb.
3. $\frac{5}{7} \div 2 = \frac{5}{14}$ — Ans. $\frac{5}{14}$ kg
4. $\frac{5}{6} \div 4 = \frac{5}{24}$ — Ans. $\frac{5}{24}$ m
5. $\frac{4}{5} \div 3 = \frac{4}{15}$ — Ans. $\frac{4}{15}$ L
6. $\frac{2}{5} \div 3 = \frac{2}{15}$ — Ans. $\frac{2}{15}$ lb.
7. $\frac{8}{9} \div 4 = \frac{2}{9}$ — Ans. $\frac{2}{9}$ L
8. $\frac{6}{7} \div 3 = \frac{2}{7}$ — Ans. $\frac{2}{7}$ lb.
9. $\frac{5}{7} \div 5 = \frac{1}{7}$ — Ans. $\frac{1}{7}$ kg
10. $\frac{8}{9} \div 6 = \frac{4}{27}$ — Ans. $\frac{4}{27}$ m

7 Fractions
pp 14, 15

1. $5 \times 4 = 20$ — Ans. 20 L
2. $7 \times \frac{5}{6} = \frac{35}{6}$ — Ans. $\frac{35}{6}$ L $\left(5\frac{5}{6}\text{ L}\right)$
3. $6 \times \frac{3}{5} = \frac{18}{5}$ — Ans. $\frac{18}{5}$ m² $\left(3\frac{3}{5}\text{ m}^2\right)$
4. $80 \times \frac{3}{4} = 60$ — Ans. 60¢
5. $9 \times \frac{4}{5} = \frac{36}{5}$ — Ans. $\frac{36}{5}$ L $\left(7\frac{1}{5}\text{ L}\right)$
6. $\frac{2}{3} \times \frac{4}{5} = \frac{8}{15}$ — Ans. $\frac{8}{15}$ lb.
7. $\frac{1}{8} \times \frac{3}{7} = \frac{3}{56}$ — Ans. $\frac{3}{56}$ kg
8. $\frac{1}{5} \times \frac{3}{4} = \frac{3}{20}$ — Ans. $\frac{3}{20}$ yd.²
9. $\frac{4}{5} \times \frac{1}{2} = \frac{2}{5}$ — Ans. $\frac{2}{5}$ yd.²
10. $\frac{3}{5} \times \frac{2}{3} = \frac{2}{5}$ — Ans. $\frac{2}{5}$ yd.²

8 Fractions
pp 16, 17

1. $\frac{5}{6} \times 2 = \frac{5}{3}$ — Ans. $\frac{5}{3}$ L $\left(1\frac{2}{3}\text{ L}\right)$
2. $2 \times \frac{3}{4} = \frac{3}{2}$ — Ans. $\frac{3}{2}$ lb. $\left(1\frac{1}{2}\text{ lb.}\right)$
3. $36 \times \frac{2}{3} = 24$ — Ans. 24 yd.
4. $\frac{3}{4} \times \frac{5}{6} = \frac{5}{8}$ — Ans. $\frac{5}{8}$ m
5. $\frac{4}{5} \times \frac{2}{3} = \frac{8}{15}$ — Ans. $\frac{8}{15}$ kg
6. $\frac{7}{8} \times \frac{3}{4} = \frac{21}{32}$ — Ans. $\frac{21}{32}$ L
7. $360 \times \frac{3}{5} = 216$ — Ans. 216 students
8. $\frac{6}{7} \times \frac{2}{3} = \frac{4}{7}$ — Ans. $\frac{4}{7}$ kg
9. $\frac{9}{10} \times \frac{1}{3} = \frac{3}{10}$ — Ans. $\frac{3}{10}$ lb.
10. $\frac{8}{5} \times \frac{5}{6} = \frac{4}{3}$ — Ans. $\frac{4}{3}$ ft. $\left(1\frac{1}{3}\text{ ft.}\right)$

9 Fractions
pp 18, 19

1. $10 \div 3 = \frac{10}{3}$ — Ans. $\frac{10}{3}$ L $\left(3\frac{1}{3}\text{ L}\right)$
2. $10 \div \frac{3}{5} = \frac{50}{3}$ — Ans. $\frac{50}{3}$ L $\left(16\frac{2}{3}\text{ L}\right)$
3. $2 \div \frac{1}{3} = 6$ — Ans. 6 brownies
4. $8 \div \frac{4}{5} = 10$ — Ans. $10
5. $450 \div \frac{3}{5} = 750$ — Ans. 750 cookies
6. $\frac{5}{8} \div 2 = \frac{5}{16}$ — Ans. $\frac{5}{16}$ lb.
7. $\frac{3}{8} \div \frac{2}{5} = \frac{15}{16}$ — Ans. $\frac{15}{16}$ kg
8. $\frac{2}{3} \div \frac{5}{6} = \frac{4}{5}$ — Ans. $\frac{4}{5}$ shirt
9. $\frac{7}{10} \div \frac{2}{3} = \frac{21}{20}$ — Ans. $\frac{21}{20}$ kg $\left(1\frac{1}{20}\text{ kg}\right)$
10. $\frac{5}{6} \div \frac{2}{7} = \frac{35}{12}$ — Ans. $\frac{35}{12}$ yd.² $\left(2\frac{11}{12}\text{ yd.}^2\right)$

10 Fractions
pp 20, 21

1. $12 \div 2 = 6$ — Ans. 6 pieces
2. $7 \div \frac{1}{6} = 42$ — Ans. 42 pieces
3. $5 \div \frac{1}{4} = 20$ — Ans. 20 people
4. $6 \div \frac{6}{5} = 5$ — Ans. 5 pieces
5. $4 \div \frac{2}{3} = 6$ — Ans. 6 cups
6. $\frac{3}{4} \div \frac{1}{12} = 9$ — Ans. 9 bags
7. $\frac{5}{6} \div \frac{5}{12} = 2$ — Ans. 2 pieces
8. $\frac{2}{3} \div \frac{2}{15} = 5$ — Ans. 5 pieces
9. $\frac{4}{5} \div \frac{4}{25} = 5$ — Ans. 5 cups
10. $\frac{5}{4} \div \frac{5}{8} = 2$ — Ans. 2 bags

11 Fractions pp 22, 23

1. $8 \div 2 = 4$ — **Ans.** 4 times
2. $1 \div \frac{3}{5} = \frac{5}{3}$ — **Ans.** $\frac{5}{3}$ times $\left(1\frac{2}{3}\text{ times}\right)$
3. $4 \div \frac{4}{5} = 5$ — **Ans.** 5 times
4. $8 \div \frac{8}{9} = 9$ — **Ans.** 9 times
5. $14 \div \frac{7}{8} = 16$ — **Ans.** 16 times
6. $\frac{4}{5} \div \frac{3}{5} = \frac{4}{3}$ — **Ans.** $\frac{4}{3}$ times $\left(1\frac{1}{3}\text{ times}\right)$
7. $\frac{6}{7} \div \frac{15}{14} = \frac{4}{5}$ — **Ans.** $\frac{4}{5}$ times
8. $\frac{2}{3} \div \frac{4}{5} = \frac{5}{6}$ — **Ans.** $\frac{5}{6}$ times
9. $\frac{3}{8} \div \frac{9}{8} = \frac{1}{3}$ — **Ans.** $\frac{1}{3}$ times
10. $\frac{7}{9} \div \frac{2}{3} = \frac{7}{6}$ — **Ans.** $\frac{7}{6}$ times $\left(1\frac{1}{6}\text{ times}\right)$

12 Fractions pp 24, 25

1. $5 - 3 \times \frac{4}{5} = \frac{13}{5}$ — **Ans.** $\frac{13}{5}$ m $\left(2\frac{3}{5}\text{ m}\right)$
2. $5 - 2 \times \frac{3}{4} = \frac{7}{2}$ — **Ans.** $\frac{7}{2}$ m $\left(3\frac{1}{2}\text{ m}\right)$
3. $5 - \frac{5}{6} \times 4 = \frac{5}{3}$ — **Ans.** $\frac{5}{3}$ ft. $\left(1\frac{2}{3}\text{ ft.}\right)$
4. $1 - \frac{2}{5} \times 2 = \frac{1}{5}$ — **Ans.** $\frac{1}{5}$ L
5. $\frac{1}{5} \times 6 - \frac{1}{5} = 1$ — **Ans.** 1 lb.
   $\left(\text{Also, } \frac{1}{5} \times (6-1) = 1\right)$
6. $4 \times \frac{7}{10} + 6 \times \frac{3}{5} = \frac{32}{5}$ — **Ans.** $\frac{32}{5}$ lb. $\left(6\frac{2}{5}\text{ lb.}\right)$
7. $5 \times \frac{3}{4} + 12 \times \frac{5}{16} = \frac{15}{2}$ — **Ans.** $\frac{15}{2}$ lb. $\left(7\frac{1}{2}\text{ lb.}\right)$
8. $\frac{3}{4} \times 20 + \frac{3}{5} \times 10 = 21$ — **Ans.** 21 lb.
9. $(180 + 240) \times \frac{5}{6} = 350$ — **Ans.** 350 cm
   $\left(\text{Also, } 180 \times \frac{5}{6} + 240 \times \frac{5}{6} = 350\right)$
10. $\frac{5}{6} \times \left(\frac{4}{5} + \frac{3}{5}\right) = \frac{7}{6}$ — **Ans.** $\frac{7}{6}$ ft.^{2} $\left(1\frac{1}{6}\text{ ft.}^2\right)$

13 Ratios pp 26, 27

1. $\frac{3}{4} \div \frac{2}{3} = \frac{9}{8}$ — **Ans.** $\frac{9}{8}$ times $\left(1\frac{1}{8}\text{ times}\right)$
2. $\frac{3}{4} \div \frac{2}{3} = \frac{9}{8}$ — **Ans.** $\frac{9}{8}$
3. $\frac{8}{9} \div \frac{1}{2} = \frac{16}{9}$ — **Ans.** $\frac{16}{9}$
4. $\left(\frac{4}{5} - \frac{2}{5}\right) \div \frac{4}{5} = \frac{1}{2}$ — **Ans.** $\frac{1}{2}$
5. $\left(\frac{5}{6} - \frac{1}{6}\right) \div \frac{5}{6} = \frac{4}{5}$ — **Ans.** $\frac{4}{5}$
6. $\left(\frac{3}{4} - \frac{1}{4}\right) \div \frac{3}{4} = \frac{2}{3}$ — **Ans.** $\frac{2}{3}$
7. $\frac{2}{5} \div \left(\frac{2}{5} + \frac{1}{3}\right) = \frac{6}{11}$ — **Ans.** $\frac{6}{11}$
8. $\frac{1}{2} \div \left(\frac{1}{2} + \frac{1}{3}\right) = \frac{3}{5}$ — **Ans.** $\frac{3}{5}$
9. $\frac{1}{4} \div \left(\frac{3}{5} + \frac{1}{4}\right) = \frac{5}{17}$ — **Ans.** $\frac{5}{17}$
10. $\frac{1}{4} \div \left(\frac{3}{8} + \frac{1}{4}\right) = \frac{2}{5}$ — **Ans.** $\frac{2}{5}$

14 Ratios pp 28, 29

1. $120 \times \frac{3}{4} = 90$ — **Ans.** 90 men
2. $2 \times \frac{3}{4} = \frac{3}{2}$ — **Ans.** $\frac{3}{2}$ L $\left(1\frac{1}{2}\text{ L}\right)$
3. $240 \times \frac{3}{8} = 90$ — **Ans.** 90 pages
4. $\frac{3}{4} \times \frac{2}{3} = \frac{1}{2}$ — **Ans.** $\frac{1}{2}$ L
5. $\frac{4}{5} \times \frac{1}{2} = \frac{2}{5}$ — **Ans.** $\frac{2}{5}$ yd.
6. $90 \div \frac{3}{4} = 120$ — **Ans.** 120 students
7. $3 \div \frac{3}{4} = 4$ — **Ans.** 4 pies
8. $144 \div \frac{9}{10} = 160$ — **Ans.** 160 cm
9. $\frac{4}{7} \div \frac{4}{5} = \frac{5}{7}$ — **Ans.** $\frac{5}{7}$ lb.
10. $\frac{8}{9} \div \frac{2}{3} = \frac{4}{3}$ — **Ans.** $\frac{4}{3}$ L $\left(1\frac{1}{3}\text{ L}\right)$

15 Average pp 30, 31

1. $(3+5+2+6) \div 4 = 4$ — **Ans.** 4 books
2. $(3.1+2.9+3.2+3+2.8) \div 5 = 3$ — **Ans.** 3 oz.
3. $(75 \times 2 + 80 \times 2 + 94) \div 5 = 80.8$ — **Ans.** 80.8
4. $(64.8+57.6+77.4) \div 3 = 66.6$ — **Ans.** 66.6 lb.
5. $(136+140+145+132+141) \div 5 = 13\overset{9}{8}.\cancel{8}$ — **Ans.** 139 cm
6. $(2+1+0+3+1) \div 5 = 1.4$ — **Ans.** 1.4 students
7. $(4+3+1+0+4+4+2) \div 7 = 2.\overset{6}{5}\cancel{7}$ — **Ans.** 2.6 eggs
8. $(63.5+62.8+63.4+63.6) \div 4 = 63.3\cancel{2}$ — **Ans.** 63.3 m
9. $(5 \times 3 + 4 \times 4) \div (3+4) = 4.4\cancel{2}$ — **Ans.** 4.4 trucks

16 Average pp 32, 33

1. $4 \times 6 = 24$ — **Ans.** 24 books

2 $36 \times 7 = 252$ **Ans.** 252 pages

3 $4.5 \times 4 = 18$ **Ans.** 18 fish

4 $0.8 \times 20 = 16$ **Ans.** 16 people

5 $0.65 \times 600 = 390$ **Ans.** 390 yd.

6 (1) $75 \times 3 = 225$ **Ans.** 225 points
(2) $225 + 95 = 320$ **Ans.** 320 points
(3) $320 \div 4 = 80$ **Ans.** 80 points

7 $(4.6 \times 4 + 4.1) \div 5 = 4.5$ **Ans.** 4.5 ft.

8 (1) $90 \times 5 = 450$ **Ans.** 450 points
(2) $88 \times 4 = 352$ **Ans.** 352 points
(3) $450 - 352 = 98$ **Ans.** 98 points

9 $62 \times 6 - 62.5 \times 5 = 59.5$ **Ans.** 59.5 g

17 Quantity per Unit

pp 34, 35

1 (1) B (2) B
(3) (A) $6 \div 5 = 1.2$ **Ans.** 1.2 chickens
(C) $8 \div 6 = 1.33\cdots$ **Ans.** 1.3 chickens
(4) Larger number
(5) (A) $5 \div 6 = 0.833\cdots$ **Ans.** 0.83 m^2
(C) $6 \div 8 = 0.75$ **Ans.** 0.75 m^2
(6) Smaller area (7) C

2 (1) (Park A) $56 \div 140 = 0.4$ **Ans.** 0.4 people
(Park B) $90 \div 200 = 0.45$ **Ans.** 0.45 people
(2) (Park A) $140 \div 56 = 2.5$ **Ans.** 2.5 m^2
(Park B) $200 \div 90 = 2.22\cdots$ **Ans.** 2.2 m^2
(3) Park B

3 (1) (Town A) $820{,}000 \div 4{,}100 = 200$ **Ans.** 200 people/km^2
(Town B) $1{,}490{,}000 \div 5{,}700 = 261.4\cdots$
Ans. 261 people/km^2
(2) Town B

18 Quantity per Unit

pp 36, 37

1 (Harris farm) $90 \div 50 = 1.8$
(Wood farm) $57 \div 30 = 1.9$ **Ans.** Wood farm

2 (Car A) $315 \div 35 = 9$ (Car B) $380 \div 40 = 9.5$
Ans. Car B

3 (Red potato) $3 \div 1.5 = 2$
(Regular potato) $4 \div 1.6 = 2.5$ **Ans.** Regular potato

4 (Red ribbon) $1.5 \div 6 = 0.25$
(White ribbon) $1.3 \div 5 = 0.26$ **Ans.** Red ribbon

5 (Green notebook) $6.25 \div 5 = 1.25$
(Brown notebook) $5.20 \div 4 = 1.30$ **Ans.** Green notebooks

6 (Ryan's carrots) $360 \div 40 = 9$
(Kevin's carrots) $460 \div 50 = 9.2$ **Ans.** Kevin's field

7 (Park) $40 \div 500 = 0.08$
(School) $30 \div 300 = 0.1$ **Ans.** School

8 (School A) $820 \div 600 = 1.36\cdots$
(School B) $782 \div 580 = 1.34\cdots$ **Ans.** School A

9 $93{,}080{,}000 \div 231{,}000 = 402.94\cdots$ **Ans.** 402.9 people/km^2

10 (My town) $7{,}824 \div 38 = 205.8\cdots$
(My aunt's town) $9{,}240 \div 42 = 220$
Ans. My aunt's town

19 Speed

pp 38, 39

1 $80 \div 2 = 40$ **Ans.** 40 mph

2 $2{,}700 \div 15 = 180$ **Ans.** 180 ft./min

3 $120 \div 8 = 15$ **Ans.** 15 m/sec

4 $4{,}000 \div 25 = 160$ **Ans.** 160 ft./min

5 $200 \div 2.5 = 80$ **Ans.** 80 mph

6 $90 \div \frac{3}{4} = 120$ **Ans.** 120 km/h

7 $90 \div \frac{45}{60} = 120$ **Ans.** 120 km/h

8 $24 \div \frac{40}{60} = 36$ **Ans.** 36 km/h

9 $4 \div \frac{20}{60} = 12$ **Ans.** 12 km/h

10 $150 \div \frac{24}{60} = 375$ **Ans.** 375 m/min

20 Speed

pp 40, 41

1 $40 \times 3 = 120$ **Ans.** 120 mi.

2 $64 \times 2.5 = 160$ **Ans.** 160 km

3 $300 \times 15 = 4{,}500$ **Ans.** 4,500 yd.

4 $500 \times 25 = 12{,}500$ **Ans.** 12,500 yd.

5 $5 \times 60 = 300,\ 300 \times 5 = 1{,}500$ **Ans.** 1,500 mi.

6 $60 \times \frac{45}{60} = 45$ **Ans.** 45 mi.

7 $13 \times \frac{50}{60} = \frac{65}{6}$ **Ans.** $\frac{65}{6}$ mi. $\left(10\frac{5}{6}\text{ mi.}\right)$

8 1 hour 45 minutes = 105 minutes = $\frac{105}{60}$ hours
$80 \times \frac{105}{60} = 140$ **Ans.** 140 km

9 1 hour 24 minutes = 84 minutes = $\frac{84}{60}$ hours
$40 \times \frac{84}{60} = 56$ **Ans.** 56 mi.

10 $3 \times \frac{25}{60} = \frac{5}{4}$ **Ans.** $\frac{5}{4}$ mi. $\left(1\frac{1}{4}\text{ mi.}\right)$

21 Speed

pp 42, 43

1 $12 \div 3 = 4$ **Ans.** 4 hours

2 $910 \div 65 = 14$ **Ans.** 14 minutes

3 $5 \div 0.1 = 50$ **Ans.** 50 minutes

4 $15 \div 10 = 1.5$
Ans. 1.5 hours (1 hour 30 minutes, 90 minutes)

5 $54 \div 0.5 = 108$, 108 minutes = 1 hour 48 minutes

Ans. 1 hour 48 minutes (108 minutes)

6 $34 \div 40 = \frac{17}{20}$, $\frac{17}{20}$ hour = 51 minutes **Ans.** 51 minutes

7 $880 \div 480 = \frac{11}{6} = 1\frac{5}{6}$

$1\frac{5}{6}$ hours = 1 hour 50 minutes

Ans. 1 hour 50 minutes (110 minutes)

8 $6 \div \frac{3}{14} = 28$ **Ans.** 28 minutes

9 $2 \div 3 = \frac{2}{3}$, $\frac{2}{3}$ hour = 40 minutes **Ans.** 40 minutes

10 $11 \div 15 = \frac{11}{15}$, $\frac{11}{15}$ hour = 44 minutes **Ans.** 44 minutes

22 Common Factors

pp 44, 45

1 **Ans.** 12 in.

2 **Ans.** 18 in.

3 **Ans.** 24 seconds

4 **Ans.** 7:24 a.m.

5 144 seconds = 2 minutes 24 seconds

Ans. 2 minutes 24 seconds

6 (1) **Ans.** 18 cm

(2) $(18 \div 6) \times (18 \div 9) = 6$ **Ans.** 6 cards

7 $(24 \div 8) \times (24 \div 12) = 6$ **Ans.** 6 cards

8 (1) $(24 \div 6) \times (24 \div 8) = 12$ **Ans.** 12 cards

(2) $(48 \div 6) \times (48 \div 8) = 48$ **Ans.** 48 cards

9 $(24 \div 6) \times (24 \div 8) \times (24 \div 3) = 96$ **Ans.** 96 blocks

23 Common Factors

pp 46, 47

1 B 4 C 3 D 2

2 (1) 2 (2) 4

3 The common factors of 9 and 6 are 1 and 3. **Ans.** 3 people

4 The common factors of 8 and 12 are 1, 2 and 4.

Ans. 2 people, 4 people

5 The common factors of 16 and 24 are 1, 2, 4 and 8.

Ans. 2 people, 4 people, 8 people

6 G.C.F. of 24 and 32 is 8. **Ans.** 8 groups

7 G.C.F. of 24 and 30 is 6. $24 \div 6 = 4$, $30 \div 6 = 5$

Ans. 4 candies and 5 cookies

8 G.C.F. of 36 and 42 is 6. $36 \div 6 = 6$, $42 \div 6 = 7$

Ans. 6 red sheets, 7 blue sheets

9 G.C.F. of 18 and 24 is 6. **Ans.** 6 in.

24 Speed & Distance

pp 48, 49

1 $60 \div 4 = 15$ **Ans.** 15 m/sec

2 $(80 + 120) \div 5 = 40$ **Ans.** 40 m/sec

3 $(80 + 240) \div 8 = 40$ **Ans.** 40 m/sec

4 $(90 + 660) \div 30 = 25$, $25 \times 60 = 1{,}500$ **Ans.** 1,500 m/min

[Also, 30 seconds = 0.5 minute, $(90 + 660) \div 0.5 = 1{,}500$]

5 $50 \times 20 - 900 = 100$ **Ans.** 100 m

6 $20 \times 20 - 330 = 70$ **Ans.** 70 m

7 $(120 + 580) \div 25 = 28$ **Ans.** 28 seconds

8 $(145 + 453) \div 23 = 26$ **Ans.** 26 seconds

9 $1{,}800 \div 60 = 30$, 1.8 km/min = 30 m/sec

$(100 + 500) \div 30 = 20$ **Ans.** 20 seconds

10 $86{,}400 \div 60 \div 60 = 24$

86.4 km/hr = 24 m/sec

2 minutes 10 seconds = 130 seconds

$24 \times 130 - 2{,}970 = 150$ **Ans.** 150 m

25 Rate

pp 50, 51

1 $5{,}625 \div (60 + 65) = 45$ **Ans.** 45 minutes

2 $3{,}350 \div (70 + 64) = 25$ **Ans.** 25 minutes

3 $12 \div (4.2 + 4.8) = \frac{4}{3}$ $\frac{4}{3}$ hours = 1 hour 20 minutes

Ans. 1 hour 20 minutes

(Also, $4{,}200 \div 60 = 70$, 4.2 km/hr = 70 m/min,
$4{,}800 \div 60 = 80$, 4.8 km/hr = 80 m/min,
$12{,}000 \div (70 + 80) = 80$,
80 minutes = 1 hour and 20 minutes)

4 $1{,}000 \div (10 + 15) = 40$ **Ans.** 40 minutes

5 $27 \div (1.5 + 3) = 6$ **Ans.** 6 months

6 $3{,}000 \div (80 + 70) = 20$ **Ans.** 20 minutes

7 $225 \div (40 + 50) = 2.5$

Ans. 2.5 hours (2 hours 30 minutes)

8 $2{,}000 \div (250 - 125) = 16$ **Ans.** 16 minutes

9 $4.5 \div (3.5 - 2) = 3$ **Ans.** 3 months

10 $60 \times 14 \div (200 - 60) = 6$ **Ans.** 6 minutes

26 Mixed Problems

pp 52, 53

1 (1) $100 \div 5 \times 2 = 40$ **Ans.** \$40

(2) $100 \div 5 \times 3 = 60$

[Also, $100 - 40 = 60$] **Ans.** \$60

2 (Rachel) $120 \div 6 \times 4 = 80$

(Sister) $120 \div 6 \times 2 = 40$

[Also, $120 - 80 = 40$] **Ans.** (Rachel) \$80, (Sister) \$40

3

(Joanne) $140 \div 7 \times 4 = 80$

(Veronica) $140 \div 7 \times 3 = 60$

[Also, $140 - 80 = 60$]

Ans. (Joanne) 80 stickers, (Veronica) 60 stickers

4 $88 \div 11 \times 8 = 64$ **Ans.** 64 cm

5 $140 \div 10 \times 6 = 84$ **Ans.** 84 in.

27 Mixed Problems

pp 54, 55

1 (1) $(500-20) \div 3 = 160$ **Ans.** 160 gumballs

(2) $160 \times 2 + 20 = 340$

[Also, $500 - 160 = 340$] **Ans.** 340 gumballs

2 (Small box) $(140-20) \div 4 = 30$

(Big box) $30 \times 3 + 20 = 110$

[Also, $140 - 30 = 110$]

Ans. (Big box) 110 oranges, (Small box) 30 oranges

3

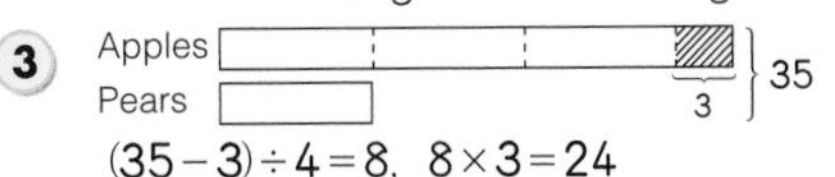

$(35-3) \div 4 = 8$, $8 \times 3 = 24$

Ans. 24 apples

4 (1) $(110+10) \div 3 = 40$ **Ans.** \$40

(2) $40 \times 2 - 10 = 70$ **Ans.** \$70

[Also, $110 - 40 = 70$]

5

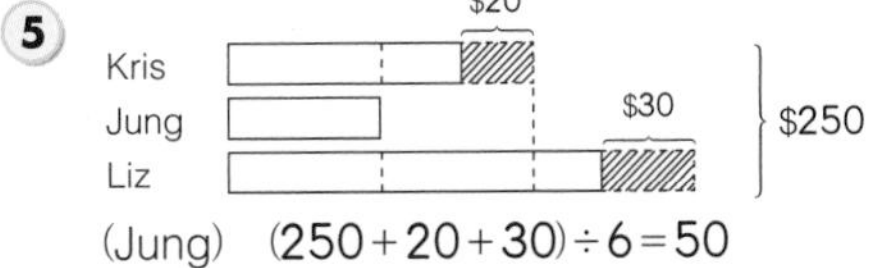

(Jung) $(250+20+30) \div 6 = 50$

(Kris) $50 \times 2 - 20 = 80$

(Liz) $50 \times 3 - 30 = 120$

Ans. (Kris) \$80, (Jung) \$50, (Liz) \$120

6

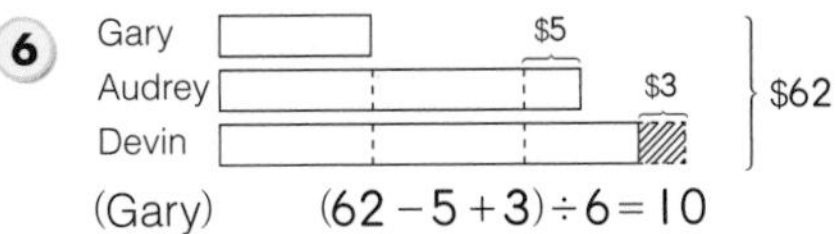

(Gary) $(62 - 5 + 3) \div 6 = 10$

(Audrey) $10 \times 2 + 5 = 25$

(Devin) $10 \times 3 - 3 = 27$

Ans. (Gary) \$10, (Audrey) \$25, (Devin) \$27

28 Mixed Problems

pp 56, 57

1 (1) (A) $\frac{1}{10}$, (B) $\frac{1}{15}$

(2) $\frac{1}{10} + \frac{1}{15} = \frac{1}{6}$ **Ans.** $\frac{1}{6}$

(3) $1 \div \frac{1}{6} = 6$ **Ans.** 6 days

2 (1) $\frac{1}{9} + \frac{1}{18} = \frac{1}{6}$ **Ans.** $\frac{1}{6}$

(2) $1 \div \frac{1}{6} = 6$ **Ans.** 6 days

3 $1 \div \left(\frac{1}{20} + \frac{1}{30}\right) = 12$ **Ans.** 12 days

4 $1 \div \left(\frac{1}{10} + \frac{1}{12} + \frac{1}{15}\right) = 4$ **Ans.** 4 days

5 (1) $\frac{1}{12}$

(2) $\frac{1}{8} - \frac{1}{12} = \frac{1}{24}$ **Ans.** $\frac{1}{24}$

(3) $1 \div \frac{1}{24} = 24$ **Ans.** 24 days

6 $\frac{1}{9} - \frac{1}{12} = \frac{1}{36}$, $1 \div \frac{1}{36} = 36$ **Ans.** 36 days

7 $\frac{1}{15} - \frac{1}{20} = \frac{1}{60}$, $1 \div \frac{1}{60} = 60$ **Ans.** 60 days

29 Mixed Problems

pp 58, 59

1 (1) $10 \times 20 - 130 = 70$ **Ans.** 70¢

(2) $10 - 5 = 5$ **Ans.** 5¢

(3) $70 \div 5 = 14$ **Ans.** 14 dimes

(4) $20 - 14 = 6$ **Ans.** 6 dimes

2 (1) $4 \times 10 - 26 = 14$ **Ans.** 14 legs

(2) $4 - 2 = 2$, $14 \div 2 = 7$ **Ans.** 7 turtles

(3) $10 - 7 = 3$ **Ans.** 3 turtles

3 (1) $(50 \times 12 - 450) \div (50 - 20) = 5$ **Ans.** 5 stamps

(2) $12 - 5 = 7$ **Ans.** 7 stamps

4 (1) $(1 \times 12 - 10) \div (1 - 0.8) = 10$ **Ans.** 10 pencils

(2) $12 - 10 = 2$ **Ans.** 2 pencils

5 (1) $(8 \times 10 - 65) \div (8 - 5) = 5$ **Ans.** 5 sticks

(2) $10 - 5 = 5$ **Ans.** 5 sticks

6 $(0.7 \times 16 - 9) \div (0.7 - 0.5) = 11$ **Ans.** 11 gums

30 Mixed Problems

pp 60, 61

1 (1) $12 \div 2 = 6$ **Ans.** 6 students

(2) $3 \times 6 = 18$ **Ans.** 18 pencils

[Also, $5 \times 6 - 12 = 18$]

2 (1) $15 \div (7-4) = 5$ **Ans.** 5 students

(2) $4 \times 5 = 20$ **Ans.** 20 paper clips

[Also, $7 \times 5 - 15 = 20$]

3 $18 \div (10-7) = 6$, $7 \times 6 = 42$

Ans. 6 children, 42 candies

[Also, $18 \div (10-7) = 6$, $10 \times 6 - 18 = 42$]

4 (1) $12 \div 2 = 6$ **Ans.** 6 children

(2) $5 \times 6 = 30$ **Ans.** 30 grapes

[Also, $3 \times 6 + 12 = 30$]

5 (1) $16 \div (8-6) = 8$ **Ans.** 8 students

(2) $8 \times 8 = 64$ **Ans.** 64 sheets of paper

[Also, $6 \times 8 + 16 = 64$]

6 $(4+6) \div (7-5) = 5$ **Ans.** 5 children

7 $(11+9) \div (7-5) = 10$ **Ans.** 10 students

31 Proportion

pp 62, 63

1 (1) ✓ (2) ✕ (3) ✕ (4) ✓ (5) ✓ (6) ✕

2 (1)

Time (min.)	1	2	3	4	5	6	7	…
Volume of water (L)	3	6	9	12	15	18	21	…

(✓)

(2)

Mother's age	25	26	27	28	29	30	31	…
Child's age	1	2	3	4	5	6	7	…

(×)

(3)

Number of loaves	1	2	3	4	5	6	7	…
Costs ($)	0.70	1.40	2.10	2.80	3.50	4.20	4.90	…

(✓)

(4)

Number of sides	3	4	5	6	7	8	9	…
Sum of interior angles (°)	180	360	540	720	900	1080	1260	…

(×)

(5)

Time (hours)	1	2	3	4	5	6	7	…
Distance (km)	80	160	240	320	400	480	560	…

(✓)

32 Proportion

pp 64, 65

1 (1) $1 \div 3 = \frac{1}{3}$ **Ans.** $\frac{1}{3}$ (2) $3 \div 9 = \frac{1}{3}$ **Ans.** $\frac{1}{3}$

(3) $4 \div 8 = \frac{1}{2}$ **Ans.** $\frac{1}{2}$ (4) $12 \div 24 = \frac{1}{2}$ **Ans.** $\frac{1}{2}$

2 (1) (Time) $2 \div 5 = \frac{2}{5}$ **Ans.** $\frac{2}{5}$

(Depth) $8 \div 20 = \frac{2}{5}$ **Ans.** $\frac{2}{5}$

(2) (Time) $6 \div 9 = \frac{2}{3}$ **Ans.** $\frac{2}{3}$

(Depth) $24 \div 36 = \frac{2}{3}$ **Ans.** $\frac{2}{3}$

3 (1) 2 (2) 30 (3) 6 (4) 40

4 (1) $Y = 2 \times X$ (2) × (3) $Y = 4 \times X$

33 Proportion

pp 66, 67

1

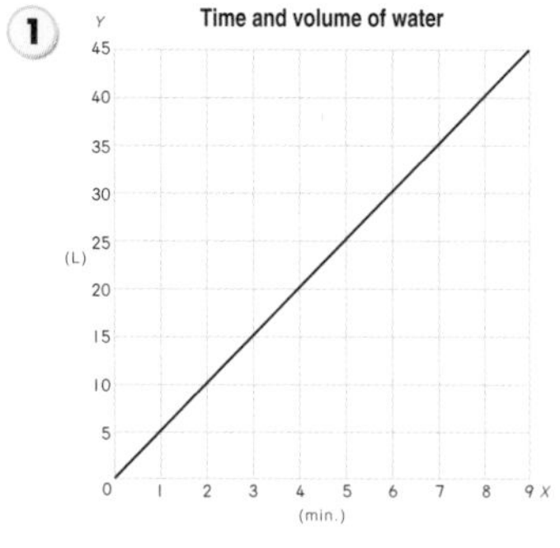

2

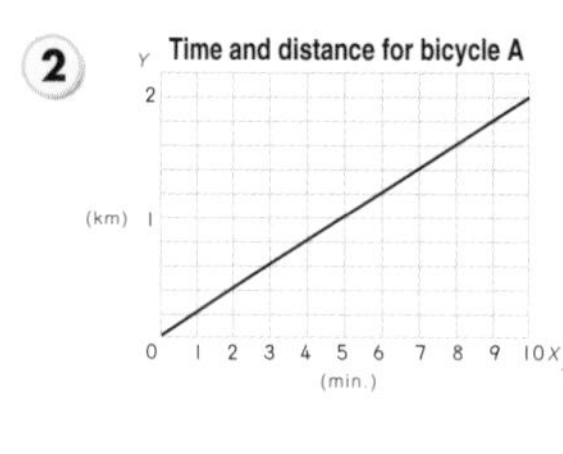

3 Length and cost of purple ribbon

Length X (m)	0	1	2	3	4	5
Cost Y ($)	0	3	6	9	12	15

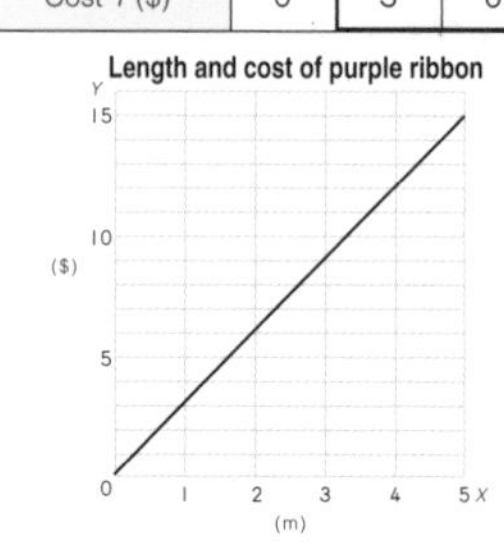

34 Proportion

pp 68, 69

1 (1) Yes (2) 30 cm (3) 4 minutes (4) 5 cm

2 (1) Yes (2) 300 g (3) 2 m (4) 75 g

3 (1)

Length X (m)	1	2	3	4	5	6
Cost Y ($)	1.20	2.40	3.60	4.80	6.00	7.20

(2) 1.2 (3) $Y = 1.2 \times X$

4 (1)

Time X (hr.)	1	2	3	4	5	6
Distance Y (km)	60	120	180	240	300	360

(2) 60 (3) $Y = 60 \times X$

35 Review

pp 70, 71

1 L.C.M. of 3 and 5 is 15. **Ans.** 15 in.

2 $\frac{3}{5} + \frac{1}{6} = \frac{23}{30}$ **Ans.** $\frac{23}{30}$ L

3 $(63+56+60+58+68) \div 5 = 61$ **Ans.** 61 g

4 $420 \div 35 = 12$ **Ans.** 12 mi.

5 $150 \times 3 = 450$ **Ans.** 450 mi.

6 $\frac{3}{5} \times 4 = \frac{12}{5}$ **Ans.** $\frac{12}{5}$ L $\left(2\frac{2}{5}\text{ L}\right)$

7 $\frac{3}{5} \div \frac{1}{3} = \frac{9}{5}$ **Ans.** $\frac{9}{5}$ m² $\left(1\frac{4}{5}\text{ m}^2\right)$

8

Time (min.)	1	2	3	4	5	6	7	…
Volume (L)	4	8	12	16	20	24	28	…

9 $\frac{2}{5} \div \frac{1}{3} = \frac{6}{5}$ **Ans.** $\frac{6}{5}$ kg $\left(1\frac{1}{5}\text{ kg}\right)$

10 $5 \times \frac{7}{10} = \frac{7}{2}$ **Ans.** $\frac{7}{2}$ g $\left(3\frac{1}{2}\text{ g}\right)$

36 Review

pp 72, 73

1 The common factors of 8 and 12 are 1, 2 and 4.

Ans. 2 children, 4 children

2 $\frac{3}{5} - \frac{1}{7} = \frac{16}{35}$ **Ans.** $\frac{16}{35}$ L

3 $1.5 \times 30 = 45$ **Ans.** 45 lb.

4 3 hours = 180 minutes, $162 \div 180 = 0.9$

Ans. 0.9 mi.

(Also, $162 \div 3 = 54$, $54 \div 60 = 0.9$)

5 $48 \div 2 = 24$ **Ans.** 24 minutes

6 $\frac{3}{4} \div 2 = \frac{3}{8}$ **Ans.** $\frac{3}{8}$ L

7 $\frac{2}{3} \div \frac{3}{5} = \frac{10}{9}$ **Ans.** $\frac{10}{9}$ $\left(1\frac{1}{9}\right)$

8 (Anne) $180 \div 9 \times 5 = 100$

(Betty) $180 \div 9 \times 4 = 80$

Ans. (Anne) 100 stamps, (Betty) 80 stamps

9 $30 \times (70 \div 10) = 210$ **Ans.** 210 g

10 $1{,}600{,}000 \div 6{,}500 = 246.\cdots$ (50) **Ans.** 250 people/km²